高等职业教育食品类专业教材

食品专业英语

杨雅兰　唐秋实　阳志军　主编
　　　　　　　杨越兰　主审

图书在版编目（CIP）数据

食品专业英语/杨雅兰，唐秋实，阳志军主编．—北京：中国轻工业出版社，2025.5

ISBN 978-7-5184-2887-8

Ⅰ.①食⋯　Ⅱ.①杨⋯ ②唐⋯ ③阳⋯　Ⅲ.①食品工业-英语-高等职业教育-教材　Ⅳ.①TS2

中国版本图书馆CIP数据核字（2020）第020539号

责任编辑：张　靓　　责任终审：白　洁　　封面设计：锋尚设计
版式设计：王超男　　责任校对：吴大鹏　　责任监印：张　可

出版发行：中国轻工业出版社（北京鲁谷东街5号，邮编：100040）
印　　刷：三河市万龙印装有限公司
经　　销：各地新华书店
版　　次：2025年5月第1版第5次印刷
开　　本：720×1000　1/16　印张：14.75
字　　数：280千字
书　　号：ISBN 978-7-5184-2887-8　定价：38.00元
邮购电话：010-85119873
发行电话：010-85119832　010-85119912
网　　址：http://www.chlip.com.cn
Email：club@chlip.com.cn
版权所有　侵权必究
如发现图书残缺请与我社邮购联系调换
250812J2C105ZBQ

《食品专业英语》编写人员

主　编　　顺德职业技术学院　　　　　　　杨雅兰
　　　　　　顺德职业技术学院　　　　　　　唐秋实
　　　　　　广东中昊药业有限公司　　　　　阳志军

副主编　　广东岭南职业技术学院　　　　　聂健
　　　　　　广东产品质量监督检验研究院　　叶翠平
　　　　　　江苏财经职业技术学院　　　　　董静

参　编　　常州工程职业技术学院　　　　　李颖超
　　　　　　连云港师范高等专科学校　　　　贾润红
　　　　　　顺德职业技术学院　　　　　　　邓志锋

前言

随着经济的飞速发展，我国在国际贸易中的地位日益重要，食品行业对专业英语人才的需求也不断增加，为适应新形势下高等职业教育的教学目标，食品专业英语课程已成为高职高专院校食品类专业学生的一门必修课程，它也是培养食品行业国际化人才的一门重要课程。根据高职高专院校培养高端技能型人才的培养目标，食品专业英语课程的主要任务是引导学生掌握一定的食品专业英语词汇和专业术语，学会专业英语的一些翻译技巧，以提高专业英语文献的阅读和翻译水平，并能够独立翻译和撰写简单的食品专业英文。同时，也有助于食品专业学生学习国外先进技术、开阔国际视野。

本教材着力体现素质教育和能力本位的精神，突出培养实用性人才的特色，坚持"以应用为目的、实用为主、够用为度"的前提下，紧紧围绕食品行业发展，进一步更新理念，更新内容，更新体系，更新要求。

本教材内容主要包括食品营养素、食品加工、食品保藏、食品添加剂、食品质量与安全管理、食品检测、食品标准与法规、功能食品八个部分，几乎涵盖食品行业各个方面，每个部分又分为几个子单元进行了相关介绍，课文后面附有相关词汇、参考译文和习题，能够满足大部分读者的需求。

全书语言精练、可读性强、覆盖面广、难易适中，旨在提高食品类专业学生的专业英语文献阅读和翻译能力，便于高职院校食品类专业师生使用，食品营养与检测、食品加工技术专业学生也可选用此教材，也可供食品行业工程技术人员参考。

本教材由顺德职业技术学院杨雅兰、唐秋实，广东中昊药业有限公司阳志军担任主编，广东岭南职业技术学院聂健、广东产品质量监督检验研究院叶翠平、江苏财经职业技术学院董静任副主编，常州工程职业技术学院李颖超、连云港师范高等专科学校贾润红、顺德职业技术学院邓志锋参编。全书由唐秋实统稿，杨雅兰、邓志锋老师校稿。

由于编者编写水平有限，编写时间仓促，书中难免有错误及疏漏之处，恳请专家和读者提出宝贵意见。

编者
2019 年 7 月

目 录
CONTENTS

Unit Ⅰ
Food Nutrients
食品营养物质

Lesson 1 ·· 1
  Reading Material　Water ·· 1
  Vocabulary ·· 4
  参考译文　水 ··· 5
  Exercise ·· 7

Lesson 2 ·· 9
  Reading Material　Carbohydrates ····································· 9
  Vocabulary ··· 12
  参考译文　碳水化合物 ·· 12
  Exercise ·· 14

Lesson 3 ·· 16
  Reading Material　Lipid ··· 16
  Vocabulary ··· 19
  参考译文　脂类 ·· 19
  Exercise ·· 21

Lesson 4 ·· 22
  Reading Material　Protein ·· 22
  Vocabulary ··· 25
  参考译文　蛋白质 ··· 25
  Exercise ·· 28

Lesson 5 ·· 29
  Reading Material　Vitamin ··· 29
  Vocabulary ··· 31

参考译文　维生素 ………………………………………………… 32
　　　Exercise ……………………………………………………………… 34
　Lesson 6 …………………………………………………………………… 35
　　　Reading Material　Mineral ……………………………………… 35
　　　Vocabulary ………………………………………………………… 37
　　　参考译文　矿物质 …………………………………………………… 38
　　　Exercise ……………………………………………………………… 40

Unit Ⅱ
Food Processing
食品加工

　Lesson 7 …………………………………………………………………… 41
　　　Reading Material　Bread Making ……………………………… 41
　　　Vocabulary ………………………………………………………… 45
　　　参考译文　面包制作 ………………………………………………… 46
　　　Exercise ……………………………………………………………… 49
　Lesson 8 …………………………………………………………………… 50
　　　Reading Material　Sausage Making …………………………… 50
　　　Vocabulary ………………………………………………………… 54
　　　参考译文　香肠制作 ………………………………………………… 55
　　　Exercise ……………………………………………………………… 58
　Lesson 9 …………………………………………………………………… 59
　　　Reading Material　Yogurt Making ……………………………… 59
　　　Vocabulary ………………………………………………………… 63
　　　参考译文　酸乳制作 ………………………………………………… 63
　　　Exercise ……………………………………………………………… 67
　Lesson 10 ………………………………………………………………… 68
　　　Reading Material　Flour Milling ………………………………… 68
　　　Vocabulary ………………………………………………………… 70
　　　参考译文　面粉加工 ………………………………………………… 71
　　　Exercise ……………………………………………………………… 72
　Lesson 11 ………………………………………………………………… 74
　　　Reading Material　Rice processing ……………………………… 74
　　　Vocabulary ………………………………………………………… 77
　　　参考译文　大米加工 ………………………………………………… 77

  Exercise ·· 79
Lesson 12 ·· 80
  Reading Material Fruit juice Processing ································ 80
  Vocabulary ·· 83
  参考译文 果汁制作 ·· 84
  Exercise ·· 86

Unit Ⅲ
Food Preservation
食品保藏

Lesson 13 ·· 88
  Reading Material Food Preservation by Canning ··················· 88
  Vocabulary ·· 90
  参考译文 食品罐藏 ·· 90
  Exercise ·· 91
Lesson 14 ·· 92
  Reading Material Food Preservation by Drying and Dehydration ·········· 92
  Vocabulary ·· 93
  参考译文 食品的干燥及脱水保藏 ································· 93
  Exercise ·· 94
Lesson 15 ·· 95
  Reading Material Low Temperature Storage ·························· 95
  Vocabulary ·· 97
  参考译文 低温储存 ·· 97
  Exercise ·· 98
Lesson 16 ·· 99
  Reading Material Food Irradiation ··· 99
  Vocabulary ·· 100
  参考译文 食物辐射 ·· 101
  Exercise ·· 102

Unit Ⅳ
Food Additives
食品添加剂

Lesson 17 ·· 103

Reading Material　Overview of Food Additives ········· 103
　　Vocabulary ········· 106
　　参考译文　食品添加剂概述 ········· 106
　　Exercise ········· 108
Lesson 18 ········· 109
　　Reading Material　Preservatives ········· 109
　　Vocabulary ········· 111
　　参考译文　防腐剂 ········· 112
　　Exercise ········· 113
Lesson 19 ········· 115
　　Reading Material　Food Colorants ········· 115
　　Vocabulary ········· 118
　　参考译文　食品着色剂 ········· 118
　　Exercise ········· 120
Lesson 20 ········· 121
　　Reading Material　Antioxidants ········· 121
　　Vocabulary ········· 123
　　参考译文　抗氧化剂 ········· 123
　　Exercise ········· 125
Lesson 21 ········· 127
　　Reading Material　Flavoring agents ········· 127
　　Vocabulary ········· 130
　　参考译文　调味剂 ········· 130
　　Exercise ········· 133

Unit Ⅴ
Food Quality and Safety Management
食品质量与安全管理

Lesson 22 ········· 134
　　Reading Material　Overview of Food Safety ········· 134
　　Vocabulary ········· 137
　　参考译文　食品安全概述 ········· 137
　　Exercise ········· 139
Lesson 23 ········· 140
　　Reading Material　Food Quality Management（ISO 9000） ········· 140

  Vocabulary ··· 143

  参考译文　食品质量管理（ISO 9000） ····························· 143

  Exercise ·· 145

 Lesson 24 ··· 146

  Reading Material　GMP and SSOP ································· 146

  Vocabulary ··· 148

  参考译文　GMP 与 SSOP ··· 148

  Exercise ·· 150

 Lesson 25 ··· 151

  Reading Material　HACCP ··· 151

  Vocabulary ··· 153

  参考译文　HACCP ··· 153

  Exercise ·· 154

 Lesson 26 ··· 156

  Reading Material　ISO 22000 ······································· 156

  Vocabulary ··· 158

  参考译文　ISO 22000 ·· 159

  Exercise ·· 161

Unit Ⅵ
Food Detection
食品检测

 Lesson 27 ··· 162

  Reading Material　Food Sample Treatment ······················ 162

  Vocabulary ··· 166

  参考译文　食品样品处理 ··· 166

  Exercise ·· 169

 Lesson 28 ··· 170

  Reading Material　Food Analysis Technologies ·················· 170

  Vocabulary ··· 175

  参考译文　食品分析技术 ··· 175

  Exercise ·· 179

 Lesson 29 ··· 180

  Reading Material　Food Rapid Detection Technology ········· 180

  Vocabulary ··· 185

参考译文　食品快速检测技术 …………………………………… 186
　　Exercise …………………………………………………………… 189
Lesson 30 ………………………………………………………………… 190
　　Reading Material　Detection of Food Microorganisms ………… 190
　　Vocabulary ………………………………………………………… 193
　　参考译文　食品微生物检测 ……………………………………… 193
　　Exercise …………………………………………………………… 195

Unit Ⅶ
Food Regulation and standard
食品法规与标准

Lesson 31 ………………………………………………………………… 196
　　Reading Material　Food Regulation ……………………………… 196
　　Vocabulary ………………………………………………………… 199
　　参考译文　食品法规 ……………………………………………… 199
　　Exercise …………………………………………………………… 201
Lesson 32 ………………………………………………………………… 202
　　Reading Material　Food Standard ………………………………… 202
　　Vocabulary ………………………………………………………… 204
　　参考译文　食品标准 ……………………………………………… 204
　　Exercise …………………………………………………………… 205

Unit Ⅷ
Functional Food
功能食品

Lesson 33 ………………………………………………………………… 207
　　Reading Material　Functional Food ……………………………… 207
　　Vocabulary ………………………………………………………… 210
　　参考译文　功能性食品 …………………………………………… 210
　　Exercise …………………………………………………………… 212

Related glossary terms …………………………………………………… 213
References ………………………………………………………………… 220

Unit 1
Food Nutrients
食品营养物质

Lesson 1

Reading Material

Water

Water, with the chemical formula H_2O, is the most important thing in our lives. A major component of food is water, which can encompass anywhere from 50% in meat products to 95% in lettuce, cabbage, and tomato products. Fruits, vegetables, juices, raw meat, fish and milk belong to the high-moisture category. Bread, hard cheeses and sausages are examples of intermediate-moisture foods, while the low-moisture group includes dehydrated vegetables, grains, milk powder and nuts. Indicative values of water content in a number of food products are shown in Table 1.1.

Table 1.1　　　　　　**Typical Water Content of Some Foods**

Food	Water/%
Cucumber	95~96

续表

Food	Water/%
Tomato	93~95
Cabbage	90~92
Orange juice	86~88
Apple	85~87
Cow milk	86~87
Egg, whole	74
Pork, Chicken	68~72
Hard cheese	30~50
White bread	34
Honey	15~23
Wheat	10~13
Nut	4~7
Milk Powder	3~4

Water is as much a part of all foods as carbohydrates, fats and proteins. The functional importance of water in foods goes far beyond its mere quantitative presence in their composition. On one hand, water is essential for the good texture and appearance of fruits and vegetables. In such products, loss of water usually results in lower quality. On the other hand, water, being an essential requirement for the occurrence and support of chemical reactions and microbial growth, is often responsible for the microbial, enzymatic and chemical deterioration of food.

Moisture content in foods is important in that, which may be held as: ① Free water, ② Bound water. Free water is present in cells, and in circulating fluids of tissues as in cell sap. It contains dissolved and dispersed solutes in the cell. It is easily lost by drying the food. The bound water in foods is held by the proteins, polysaccharides and fats in the living cells. Bound water may also be absorbed on the surfaces of solids in foods. The removal of bound water from tissues is very difficult. Bound water is resistant to freezing and drying.

Water activity

It is now well established that the effect of water on the stability of foods cannot be related solely to the quantitative water content. As an example, honey containing 23% water is perfectly shelf-stable, while dehydrated potato will undergo rapid spoilage at a moisture content half as high. To explain the influence of water, a parameter that reflects both the quantity and the "effectiveness" of water is needed.

This parameter is water activity.

Water activity, A_w, is defined as the ratio of the water vapor pressure of the food to the vapor pressure of pure water at the same temperature.

$$A_w = p/p_0 \tag{1.1}$$

where p = partial pressure of water vapor of the food at temperature T and p_0 = equilibrium vapor pressure of pure water at temperature T. Typical water activities of some foods are shown in Table 1.2.

The control of water activity in foods is an important tool for extending shelf life. It is responsible for the quality of foods affected by microbiological, chemical, and physical changes. The physical properties, quantity, and quality of water within the food have a strong impact on food effectiveness, quality attributes, shelf life, textural properties, and processing. Bacterial growth does not occur at water activity levels below 0.9. With the exception of osmophilic species, the water activity limit for the growth of molds and yeasts is between 0.8 and 0.9. Most enzymatic reactions require water activity levels of 0.85 or higher. The relationship between water activity and chemical reactions (Maillard browning, lipid oxidation) exhibits more complex behavior with maxima and minima.

Table 1.2   Typical Water Activities of Selected Foods

A_w Range	Products Examples
0.95⩾	Fresh fruits and vegetables, milk, meat, fish
0.90~0.95	Semi-hard cheeses, salted fish, bread
0.85~0.90	Hard cheese, sausage, butter
0.80~0.85	Concentrated fruit juices, jelly
0.70~0.80	Jams and preserves, prunes, dry cheeses, legumes
0.50~0.70	Raisins, honey, grains
0.40~0.50	Almonds
0.20~0.40	Nonfat milk powder
<0.2	Crackers, sugar

Water function

Water has many functions in food. It can be used as a solvent, leavening agent, cooking medium, body needs, keeping quality of foods, texture consistency and so on. For example, water functions in food preparation as a dispersing medium and helps to produce a smooth texture. It helps to distribute particles of materials like starch and protein. When flour is used to thicken liquids, the particles need to be dispersed throughout the liquid phase as in a starch gel. Moisture content in foods also has a bearing on the cooking methods employed and on the cooking time. Dry foods like cereals, millets, pulses take a longer time to cook than foods with a greater moisture content, e. g. leafy vegetables. Dry foods such as pulses, are generally first soaked for a period of time before they are cooked.

Food-preservation processes have a common goal of extending the shelf life of foods to allow for storage and convenient distribution. The activity of microorganisms is the first and most dangerous limitation of shelf life. Water is essential for microorganisms that may cause food spoilage if they are present in a food that offers them favorable conditions for growth. Hence, many food preservation techniques were developed to reduce the availability or activity of water in order to eliminate the danger of microbial spoilage.

For the food industry, water is essential for processing, as a heating or a cooling medium. It may be employed in processes in the form of liquid water or in the other states of water such as ice or steam.

Almost all food-processing techniques involve the use or the modification of water in food. Freezing, drying, concentration, and emulsification processes all involve changes in the water fraction of the food. Without the presence of water, it would not be possible to achieve the physicochemical changes that occur during cooking such as the gelatinization of starch. Water is important as a solvent for dissolving small molecules to form solutions and as a dispersing medium for dispersing larger molecules to form colloidal solutions.

Our body also rely on water that is an essential nutrient next in importance to oxygen. Deprivation of water even for a few days can lead to death. An adult may need about 1.0~1.5 liters of water per day, in addition to the moisture contained in foods eaten. The amount of water needed by human body may increase in hot and dry climate and with strenuous exercise.

Vocabulary

dehydrated 脱水的,干燥的
deprivation 损失,丧失
extend 延长,延伸
vapor 水蒸气

water 水,水分
water activity 水分活度
shelf life 保存期,货架期
preservation 保藏,保存

参考译文

水

水(H_2O)是生命之源,是食品中的主要成分之一。它分布广泛,肉制品中含水量50%,生菜、卷心菜和番茄中含水量则达95%。根据食品中水分含量可分为三类:高含水量、中度含水量和低含水量食品。水果、蔬菜、果汁、生鲜肉、鱼和牛乳都属于高含水量一类,面包、硬质奶酪和腊肠则是中度含水量食品代表,低含水量食品包括脱水蔬菜、谷物、乳粉和坚果。表1.1列出了一些食品中的水分含量的参考值。

表1.1　　　　　　　　　　一些食品中水分含量

食品	水分含量/%
黄瓜	95~96
番茄	93~95
卷心菜	90~92
橙汁	86~88
苹果	85~87
牛乳	86~87
鲜蛋	74
猪肉,鸡肉	68~72
硬质奶酪	30~50
面包	34
蜂蜜	15~23
小麦	10~13
坚果	4~7
乳粉	3~4

水和碳水化合物、脂肪、蛋白质一样都是所有食品的重要组成部分。水在食品中的重要性远远超过了它在食品中的含量。一方面,水是一些水果和蔬菜良好质地和外观呈现的重要介质,在这些产品中,水的损失通常会导致质量下降。另一方面,水是引发食品中的化学反应和支持微生物生长的必要条件,常常与微生物、酶和化学变化引起食品变质有关。

食品中的含水量如此重要,它可分为:①自由水,②结合水。自由水存在于细胞间和细胞液,包括细胞里用于溶解和分散溶质的这些水,食物在干燥时很容易失去自由水。食品中的结合水主要存在于活细胞中的蛋白质、多糖和脂肪里面。结合水也可能吸附在食品的固体表面上。从组织中去除结合水是很困难的,它比较抗寒和耐热。

水分活度 A_w

现在已经确定水对食品品质稳定性的影响不能仅仅依赖含水量。例如,蜂蜜中含水量为23%就能达到很好的货架期,而脱水土豆含水量只有蜂蜜一半时就会快速腐败。为了解释水对食品品质的影响,有一个指标参数既能反映含水量也能反映食品中水的需求量。这个参数就是水分活度。

水分活度,用 A_w 表示,被定义为相同温度下食品中水的蒸气压 p 与纯水蒸气压 p_0 的比值。

$$A_w = p/p_0 \tag{1.1}$$

式中 p——温度 T 下食品中水的蒸气分压;

p_0——温度 T 下纯水的蒸气压。常见食品中典型的水分活度见表1.2。

通过控制食品中水分活度来延长货架期是一个很重要的方式,它与受微生物、化学、物理变化引起的食品质量变化息息相关。食品中水的物理性质、含量和质量对食品的有效性、品质、货架期、质地和加工处理都有很大的影响。食品中细菌的生长所需水分活度一般不会低于0.9。除了耐高渗透压的微生物外,霉菌和酵母菌生长所需的水分活度一般在0.8~0.9之间。很多酶促反应所需水分活度为0.85甚至更高。一些化学反应(如美拉德反应,脂质氧化)在其反应所需的最大水分活度和最低水分活度之间则表现得更复杂。

表1.2　　　　　　　　常见食品中典型的水分活度

水分活度(A_w)	样品
0.95≥	鲜水果、蔬菜、牛乳、肉类、鱼
0.90~0.95	半干奶酪、咸鱼、面包
0.85~0.90	硬质奶酪、腊肠、黄油
0.80~0.85	浓缩果汁、果冻
0.70~0.80	果酱、蜜饯、干梅子、干奶酪、豆类
0.50~0.70	葡萄干、蜂蜜、谷物
0.40~0.50	杏仁
0.20~0.40	脱脂乳粉
<0.2	咸饼干、白糖

水的功能

水在食品中具有多种功能。它可以用作溶剂、发酵剂、烹饪、身体所需以及维持食品质量、质地均匀等作用。在食品制作过程中,水可以作为分散介质使质地光滑。它就像淀粉和蛋白质一样使食品物料中的颗粒分散。当淀粉用于液体勾芡的时候,淀粉微粒作为淀粉凝胶被分散到整个液相中。食品中的水分含量也会影响烹饪方法以及烹饪时间。像谷类、粟类、豆类这样的干性食物比含水量多的食物如蔬菜类需要更长的烹饪时间,干性食物如谷类在烹饪前通常要先浸泡一段时间。

食品保藏过程中有个共同的目标就是延长其货架期进而能够更好的保存和分销。微生物活性是影响食品货架期的首要和最不利的因素。水是微生物生长必不可少的,如果它们存在于食品中,将为微生物生长提供有利的条件,并导致食品腐败。因此,为了消除微生物引起腐败的风险,许多降低水分可用性和活性的食品保存技术已开发出来。

在食品工业中,水作为加热和冷却的介质,在食品加工过程中也是必不可少的。它可以以液态或其他形态如冰或蒸汽的形式在加工过程被利用。

几乎所有的食品加工技术都涉及食品中水的使用或改变。冷冻、干燥、浓缩和乳化过程都涉及食品中的水分变化。如果没有水的存在,就不可能实现在烹饪过程中发生的物理化学变化,如淀粉的糊化。水作为溶解小分子形成溶液的溶剂,以及作为分散大分子形成胶体溶液的分散介质,都是很重要的。

我们的身体也依赖于水,水是仅次于氧气的重要营养物质。即使是几天的缺水也会导致死亡。除了通过食品中摄入的水分外,成人每天可能需要约1.0~1.5L的水。在炎热和干燥的气候以及剧烈的运动中,人体所需的水量还会增加。

Exercise

1. Answer questions
(1) What is water made up of?
(2) How does the water function?
(3) What is the importance of water in food preparation?
(4) Write a short note on moisture contents in foods.
2. Translation
(1) Even grain products, which don't seem watery at all, may be up to one-third water.

(2) 全世界有无数的泉水可以提供极好的饮用水。

(3) Water has many functions in food. It can be used as a solvent, leavening agent, cooking medium, body needs, keeping quality of foods, texture consistency and so on. For example, water functions in food preparation as a dispersing medium and helps to produce a smooth texture. It helps to distribute particles of materials like starch and protein.

Lesson 2

Reading Material

Carbohydrates

Carbohydrates are mainly composed of the elements carbon, hydrogen and oxygen, which are the most widely distributed and abundant organic compounds on earth. They comprise more than 90% of the dry matter of plants. They are one of the three main energy-providing nutrients, and the other two are proteins and fats. The biosynthesis of carbohydrates in plants starting from carbon dioxide and water with the help of light energy, known as photosynthesis, is the basis for the existence of all other organisms which depend on the intake of organic substances with food. The carbohydrate foods, giving the highest yields of energy per unit land cultivated, are easy to store and transport. Hence carbohydrates are the cheapest and most abundant source of energy for human beings.

Plant Carbohydrates can be found in sap, fruits, seeds, roots and tubers, and structural tissues (including cellulose, hemicellulose, pectin and gum). In animals, carbohydrates are found in milk of mammals (lactose) and as a storage reserve (glycogen) to some extent. Plants can synthesise a variety of carbohydrates by photosynthesis. The sugars, such as glucose, sucrose and polysaccharide, starch and cellulose are important examples of carbohydrates produced by photosynthesis.

Carbohydrates have the general formula $C_x(H_2O)_y$ where x and y are whole numbers. Naturally occurring carbohydrates are interested in food chemistry; especially those which have six or multiples of six carbon atoms. Familiar examples are glucose $C_6H_{12}O_6$ and sucrose $C_{12}H_{22}O_{11}$ and starch represented by $(C_6H_{10}O_5)_n$.

The following simple equation illustrates carbohydrate synthesis in plants.

$$6CO_2 + 12H_2O \xrightarrow{\text{Sunlight}} C_6H_{12}O_6 + 6O_2 + 6H_2O$$

The solar energy used in photosynthesis is stored as chemical energy in the plant. Animals, which eat plants are able to utilize the chemical energy stored in the carbohydrate molecule to meet major part of their energy needs.

Carbohydrates are common components of foods, both as natural components and as added ingredients. Their use is large in terms of both the quantities consumed and the variety of products in which they are found. They have a central role in the metabolism of animals and plants. They have many different molecular structures, sizes, and shapes and exhibit a variety of chemical and physical properties. They are amenable to both chemical and biochemical modification, and both modifications are employed commercially in improving their properties and extending their use. They are also safe (nontoxic).

Classification

Carbohydrates are classified on the basis of their molecular size into monosaccharides, disaccharides and polysaccharides.

Monosaccharides as their name indicates, (mono meaning one and saccharide meaning sugar) are the simplest of carbohydrates because they consist of a single sugar unit. More complex carbohydrates are built from the units of monosaccharides. Three monosaccharides that are of importance in food preparation are glucose, fructose and galactose.

Disaccharides contain two monosaccharides which may be alike or different, unite with the loss of a molecule of water to form a disaccharide. Likewise, disaccharides can be hydrolysed with dilute acid or by enzymes, to produce the sugars from which they are made. Sucrose, lactose and maltose are the most familiar examples of disaccharides.

Polysaccharides, as their name indicates, ("poly" means many) consist of many units of sugar. When polysaccharides are linked together to form one molecule, they may be linked together in straight long chains, or may be branched. Starch, glycogen, celluloses, hemicelluloses, gums and pectic substances are some of the polysaccharides found in plants and animals.

Sugars

Monosaccharides and disaccharides are crystalline, water-soluble and sweet. In food preparation, sugars commonly referred to fructose and glucose, and the disaccharides sucrose (the sugar we use daily), lactose and maltose.

Sugars supply energy to our body. Each gramme of sugar supplies four calories. Sugar can be metabolized quickly to meet energy needs of our body. It is mainly used as a sweetening agent in beverages such as tea, coffee, fruit drinks, puddings, pies, cakes, biscuits and frozen desserts. When used in higher concentration, sugar acts as a preservative as well as a sweetening agent, e.g., jams, jellies, sweetened condensed milk, honey, etc. Honey, which we often eat, is concentrated solution of fructose and glucose, in which small amounts of sucrose, dextrins, mineral matter, proteins (trace) and organic acids are present.

Sugar occurs in solid in nature. When the solution is concentrated, the sugar crystallizes. This principle is used in the manufacture of sugar. When sugar is heated to a temperature above the melting point, it decomposes and forms a brown mass, which is known as caramel. Caramel has a bitter taste. In some products sugar is partially caramelized to enhance the colour and flavour of the product.

The consumption of sugar is closely related to human's health. Increased sugar production has resultedin increase in consumption beyond desirable level. Sugar is bought and used as such and it is also consumed in a variety of manufactured foods. High intake of sugar is undesirable for three reasons: ①It contributes to obesity; ②It increases rate of dental decay; ③It is possibly related to increased incidence of diabetes and coronary heart disease.

Starch

Starch is a polysaccharide which upon complete hydrolysis releases glucose. Most of the starches and starchy foods used in food preparation are obtained from cereals (rice, wheat, corn, barley), roots (cassava, arrowroot) and tubers (potatoes, sweet potatoes). Starch is present in small particles known as granules. These granules are of various shapes and sizes. Starch granules present in the corn grain is of a different shape and size from that of a potato tuber and wheat (Fig. 2.1).

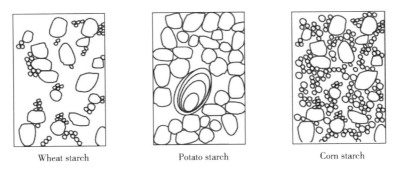

Fig. 2.1 Structure of Wheat, Potato and Corn Starch Granules

Starch is made up of two fractions amylose and amylopectin. Amylose is composed of straight-chain structure, while amylopectin has partial branched chain configuration. The two possess different properties. Amylose contributes gelling characteristics to cooked and cooled starch mixtures. Amylopectin provides cohesive or thickening property but does not usually contribute to gel formation.

In food preparation, starch is used either in the pure form (arrowroot starch, corn starch) or as cereal flour in which starch is mixed with other components (wheat flour, rice flour, corn flour).

Vocabulary

amylose 直链淀粉
amylopectin 支链淀粉
carbohydrate 碳水化合物
celluloses 纤维素
components 成分,组分
confectionaries 糖果,蜜饯
disaccharides 双糖,二糖
fructose 果糖

glucose 葡萄糖
biosynthesis 生物合成
jams 果酱
jellies 果胶,果冻
monosaccharides 单糖
sucrose 蔗糖
starch 淀粉

参考译文

碳水化合物

碳水化合物(也称糖类)主要由碳、氢、氧元素组成,是地球上分布最广泛、蕴藏量丰富的一种有机化合物,它们占植物干物质的90%以上,是供给能量的三大主要营养物质之一,另外两个是蛋白质和脂肪。植物中碳水化合物是所有其他生物体存在的基础,这些生物体依赖食品中有机物的摄入,碳水化合物的生物合成是在光照的条件下,由二氧化碳和水合成的,也就是光合作用。碳水化合物食品很容易储存和运输,它在单位土地上能源产量最高,因此,碳水化合物是人类最廉价和最丰富的能量来源。

植物中的碳水化合物存在于植物汁液、水果、种子、根和块茎以及结构组织(纤维素、半纤维素、果胶和树胶)中。动物中的碳水化合物存在于哺乳动物的奶(乳糖)中并在一定程度上作为储备的能源(糖原)。植物可以通过光合作用

合成各种碳水化合物。糖包括葡萄糖、蔗糖和多糖,还有淀粉和纤维素是光合作用生成碳水化合物的重要例子。

碳水化合物的通式为 $C_x(H_2O)_y$,其中 x 和 y 是整数。食品化学对天然存在的碳水化合物比较感兴趣,尤其是那些有六个或六个碳原子倍数的碳水化合物,常见的例子如葡萄糖($C_6H_{12}O_6$)、蔗糖($C_{12}H_{22}O_{11}$)和非还原糖淀粉$[(C_6H_{10}O_5)_n]$。

下面简单的反应式说明了植物中碳水化合物的合成。

$$6CO_2 + 12H_2O \xrightarrow{光能} C_6H_{12}O_6 + 6O_2 + 6H_2O$$

光合作用所用的太阳能以化学能的形式储存在植物中。草食动物能够利用储存在碳水化合物分子中的化学能来满足其大部分能量需求。

碳水化合物是食物的常见组分,既作为天然成分又作为添加成分。碳水化合物被需求时,其含量在消耗数量以及不同的产品种类中都不一样。它们在动植物代谢中起着核心作用。它们具有许多不同的分子结构、大小和形状,并具有各种化学和物理性质。它们适用于化学和生物化学改性,并且这两种改性在商业上都用于改善其性能并延长其用途,当然它们也是安全无毒的。

分类

根据碳水化合物分子大小可将其分为单糖、双糖和多糖。

单糖,顾名思义,就是单个的糖,由单个糖单位组成,是最简单的碳水化合物。复杂碳水化合物则由多个单糖组成。食品加工中比较重要的三种单糖是葡萄糖、果糖(左旋)和半乳糖。

双糖,由两个相同或不同的单糖通过失去一分子的水而组成。同时,双糖也可以通过稀酸或酶水解而分解成两个单糖。蔗糖、乳糖和麦芽糖是最常见的双糖。

多糖,顾名思义,由多个单糖组成,可通过直链或支链形式形成一个多糖分子。植物和动物中的一些多糖有淀粉、糖原、纤维素、半纤维素、树胶和果胶物质。

糖

单糖和双糖具备结晶性、水溶性和甜味。食品加工中常说的糖就是指果糖和葡萄糖,以及一些双糖如蔗糖(我们日常使用的糖)、乳糖和麦芽糖。

糖能够为我们身体提供能量,每克糖提供 4 卡能量。糖能迅速代谢,以满足我们身体的能量需求。它主要用作甜味剂,如茶、咖啡、果汁饮料,布丁、馅饼、蛋糕、饼干和冷冻甜点。高浓度的糖作为甜味剂时还可以作为防腐剂,如果酱、果冻、甜炼乳、蜂蜜等。我们常吃的蜂蜜就是果糖和葡萄糖的浓缩液,其中存在少量的蔗糖、糊精、矿物质、蛋白质(痕量)和有机酸。

糖在自然条件下以固体形式存在,如浓缩后糖溶液中的糖结晶,这是制糖的一种原理。当糖被加热到高于熔点的温度时，它分解并形成一种褐色物质，称为焦糖，焦糖有苦味。有些产品就是使其中部分的糖焦糖化，以改善产品的颜色和风味。

糖的消费与人的健康有着密切联系。糖产量的增加导致糖的消费远超预期水平。购买的糖除了使用外，还用于各种食品生产。人们并不喜欢摄入高糖，有三个原因：①它会导致肥胖；②患龋齿风险增加；③可能增加糖尿病和冠心病的发病率。

淀粉

淀粉是一种多糖，完全水解后为葡萄糖。食品加工中的淀粉和淀粉食品大多从谷物（水稻、小麦、玉米、大麦）、根（木薯、竹芋）和块茎（马铃薯、红薯）中获得。淀粉以小颗粒形式存在。这些颗粒具有各种形状和尺寸。存在于玉米粒中的淀粉颗粒就与马铃薯的形状和尺寸不同（图2.1）。

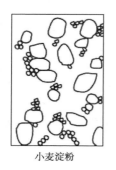 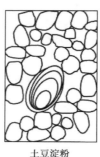 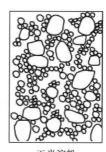

小麦淀粉　　　　　土豆淀粉　　　　　玉米淀粉

图 2.1　小麦、土豆和玉米中淀粉颗粒的结构

淀粉由直链淀粉和支链淀粉两部分组成。直链淀粉由直链结构组成，而支链淀粉具有部分支链结构，两者具有不同的特性。直链淀粉有助于淀粉在加热和冷却过程中的凝胶形成。支链淀粉则通常不会有助于凝胶形成，它提供黏性或增稠性。

在食品加工中，淀粉要么是以纯物质使用（如竹芋淀粉、玉米淀粉），要么与其他成分混合以谷物粉形式使用（如面粉、米粉、玉米粉）。

Exercise

1. Answer questions

(1) What are carbohydrates? How do they occur in nature?

(2) How are carbohydrates classified?

(3) What different sugars are found in foods?

(4) What are starches? How do they occur in nature?

2. Translation

(1) Carbohydrates are common components of foods, both as natural components and as added ingredients. Their use is large in terms of both the quantities consumed and the variety of products in which they are found. They have a central role in the metabolism of animals and plants.

(2) Starch is a polysaccharide which upon complete hydrolysis releases glucose. Most of the starches and starchy foods used in food preparation are obtained from cereals (rice, wheat, corn, barley), roots (cassava, arrowroot) and tubers (potatoes, sweet potatoes).

Lesson 3

Reading Material

Lipid

In both animal and plant foods, there are three groups of important natural organic compounds, oils and fats, carbohydrates and proteins. These are essential nutrients which sustain life. Oils and Fats have a simple molecular structure. Oils and fats belong to a naturally occurring substances called lipids.

Food lipids are either consumed in the form of "visible" fats, which have been separated from the original plant or animal sources, such as butter, lard and shortening, or as constituents of basic foods, such as milk, cheese and meat. The largest supply of vegetable oil comes from the seeds of soybean, cottonseed and peanut, and the oil-bearing trees of palm, coconut and olive. Animals secrete fat in the milk, which is extracted as cream or butter. Store fat from animals in adipose tissues can also be extracted, such as lard and butter. Most cereals, vegetables and fruits are low in fat, With the exception of corn, which contain sufficient fat to permit commercial production.

Composition and Properties

Lipids consist of a broad group of compounds that are generally soluble in organic solvents but only sparingly soluble in water. They are major components of adipose tissue, and together with proteins and carbohydrates, they constitute the principal structural components of all living cells. Some important examples of lipids which are derivatives of fatty acids are oils, fats, phospholipids and waxes. Steroids also are lipids, which are an exception in that these are not derivatives of fatty acids. Cholesterol, a kind of steroid, is an important constituent of body tissues and is present in animal foods. Vitamin D and bile acids are other important steroids, which are related to cholesterol.

Oils and fats are composed of the elements carbon, hydrogen and oxygen. Fats are built up by linking together a number of individual fatty acids with glycerol. Glycerides are formed by the combination of glycerol and fatty acid with elimination of water as shown below:

$$\begin{array}{c} CH_2OH \\ | \\ CHOH \\ | \\ CH_2OH \end{array} + RCOOH \longrightarrow \begin{array}{c} H_2-C-O-C\lessgtr^O_R \\ | \\ H-C-O-C\lessgtr^O_R \\ | \\ H_2-C-O-C\lessgtr^O_R \end{array} + 3HOH$$

Glycerol+Fatty Acid   Simple Triglyceride+water

Fatty acids are found in all simple and compound lipids. Some common fatty acids are palmitic, stearic, oleic and linoleic acid. Fatty acids differ from one another in their chain length (the number of carbon atoms in each molecule) and the degree of saturation. There are short chain fatty acids (with a chain length of 10 or fewer carbon atoms), such as acetic acid found in vinegar and caproic acid in butter. Long chain fatty acids have a chain length of 12 to 18 carbon atoms and including palmitic and stearic acid found in lard and beef tallow respectively. Oleic acid and linoleic acid (18 carbon atoms) are also long chain fatty acids. They are found in olive and corn oils respectively.

Fatty acids can be saturated (no double bond), monounsaturated (one double bond), or polyunsaturated (two or more double bonds), and areessential for energetic, metabolic, and structural activities. An unsaturated fatty acid with a double bond can have two possible configurations, either cis or trans, depending on the relative positions of the alkyl groups.

When three fatty acids in a triglyceride are of the same kind, the fat called simple triglyceride. If the fatty acids are different, the fat called mixed glyceride. Edible fats are complex mixtures of mixed triglycerides and small amount of other associated substances occurring naturally in plants and animals. This may account for the wide variation in the flavour and consistency of food fats.

Oils and fats are similar in composition, but physically, fats are solid at normal temperatures (18~25℃) as they contain a high proportion of saturated fatty acids, whereas oils are liquids as containing a high proportion of unsaturated fatty acids. Some examples of solid fats are butter, vanaspati and margarine. Vanaspati and margarine are hydrogenated fats and hence are solid at room temperature. Liquid fats better known as oils, are liquid at room temperature. Oils such as corn, soyabean, cottonseed and safflower contain a fairly large proportion of polyunsaturated fatty acids.

Lipid Function

Lipids in food exhibit unique physical and chemical properties. Their composition, crystalline structure, melting properties, and ability to associate with water and other non lipid molecules are especially important to their functional properties in many foods. Each gramme of pure oil or fat supplies nine calories in contrast to starchy foods, which provide only four calories per gramme. Oils and fats provide 10 to 30 percent of our daily energy intake. Even when no oil or fat is added to the diet, the natural fat in the foods provides 10 to 12 percent of the total energy intake.

Fats and oils have other functions in the body besides supplying energy. They carry fat soluble vitamins A, D, E and K into the body and assist in the absorption of these vitamins. Some vegetable oils contain an essential fatty acid, which is necessary for normal body functions. Essential fatty acid is not synthesized in the body. Fats impart special flavour and texture to our foods, thus they can increas the palatability. Fats are also valuable for the satiety value that they give to meals. They are slow in leaving the stomach and hence may delay the recurrence of hunger pang.

Fats and oils are also used as a medium of cooking in shallow and deep fat frying of foods. Some lipid compounds are indispensable as food emulsifiers, while others are important as fat-soluble or oil-soluble pigments or food colorants.

Lipid Oxidation

During the processing and storage of foods, lipids undergo complex chemical changes and react with other food constituents, producing numerous compounds both desirable and deleterious to food quality. Lipid oxidation of unsaturated fatty acids is the main reaction responsible for the degradation of lipids and one of the major causes of food spoilage. Indeed, the oxidation level of oil and fat is an important quality criterion for the food industry. Oxidation of oils not only produces rancid flavors but can also decrease the nutritional quality and safety by the formation of oxidation products, which may play a role in the development of diseases. The refined oil has to be stored under an inert gas such as nitrogen or vacuum packed to prevent oxidation. It can also be prevented by addition of small quantities of chemicals which prevents the oxidation of the fats. These chemicals are called "antioxidants".

Vocabulary

butter　黄油
cheese　奶酪
coconut　椰子
cereal　谷类植物
favor　风味
fat　脂肪
lard　猪油

shortening　酥油
lipid　脂质,脂类
oil　油
olive　橄榄
palatability　适口性
seed　种子
peanut　花生

参考译文

脂　类

在动物和植物源食品中,有三种重要的天然存在的有机化合物:油脂、碳水化合物和蛋白质,这些是维持生命的必需营养物质。油和脂肪分子结构简单,它们是一种天然的脂类物质。

食品中的脂类要么以"可见"的脂肪形式使用,如从植物或动物中分离出来的脂类——黄油、猪油和酥油,要么作为食品中的基本成分,如牛乳、奶酪和肉类。植物油的最大来源是大豆、棉籽和花生的种子,以及棕榈、椰子和橄榄这些油料树。藏于动物母乳中的脂肪,经提取后可制成奶油或黄油。动物脂肪组织中的脂肪也可以提炼,如猪油和牛油就是从其中提取。大部分谷物、蔬菜和水果脂肪含量低,玉米则是例外,它含有足够的脂肪,能够商业化生产。

组成和特点

脂类是一种普遍溶于有机溶剂而微溶于水的物质,它们是脂肪组织的主要组成部分,与蛋白质和碳水化合物一起构成所有活细胞的主要结构成分。一些重要的脂类如油、脂肪、磷脂和蜡都是脂肪酸衍生物。类固醇也是一种脂类,但它不是脂肪酸的衍生物。胆固醇就是一种类固醇,是身体组织的重要组成部分,存在于动物性食品中。维生素 D 及胆汁酸是另一类重要的类固醇,与胆固醇有关。

油脂是由碳、氢和氧元素组成的。一些单独的脂肪酸与丙三醇(甘油)结合形成脂肪。丙三醇和脂肪酸失去水而形成甘油酯的过程如下：

$$\begin{array}{c} CH_2OH \\ | \\ CHOH \\ | \\ CH_2OH \end{array} + RCOOH \longrightarrow \begin{array}{c} H_2-C-O-C{\begin{smallmatrix}O\\R\end{smallmatrix}} \\ | \\ H-C-O-C{\begin{smallmatrix}O\\R\end{smallmatrix}} \\ | \\ H_2-C-O-C{\begin{smallmatrix}O\\R\end{smallmatrix}} \end{array} + 3HOH$$

丙三醇+脂肪酸　　　　　甘油脂+水

所有的简单和复合脂类中都含有脂肪酸，一些常见的脂肪酸有棕榈酸、硬脂酸、油酸和亚油酸。不同的脂肪酸其链长(分子中的碳原子数目)和饱和度不同，短链脂肪酸(链中碳原子数≤10)如醋中的乙酸和黄油中的已酸，长链脂肪酸(链中碳原子数12~18个)如猪油和牛油中的棕榈酸和硬脂酸，以及橄榄油和玉米油中的油酸和亚油酸(18个碳原子)是长链脂肪酸。

脂肪酸可以分为饱和的(没有双键)、单不饱和的(一个双键)或多不饱和的(两个或两个以上的双键)脂肪酸，它们在能量供给、新陈代谢和结构活动中必不可少。含有双键的不饱和脂肪酸有顺式和反式两种可能的构型，这取决于烷基的相对位置。

当甘油三酯上的三个脂肪酸相同，则称为单纯甘油酯，如果不同，则称为混合甘油酯。食用油脂是一些混合甘油三酯和动植物中少量其他相关天然物质的复杂混合物，这可能是食用油脂风味和稠度变化范围大的原因。

油和脂肪在组成上相同，但物体特性上，脂肪因含有高比例的饱和脂肪酸而在常温(18~25℃)下呈固态，油则因含有高比例的不饱和脂肪酸而呈液态。黄油、人造黄油和人造奶油即是固体脂肪，人造黄油和人造奶油是氢化了的脂肪，因此室温下是固体。液体脂肪俗称油，室温下为液体，如玉米、大豆、棉籽和红花油均含有相当比例的多不饱和脂肪酸。

脂的功能

食品中的脂类具有独特的物理和化学性质。它们的组成、晶体结构、熔化特性以及与水和其他非脂质分子结合的能力对它们在各种食品中的功能特性尤其重要。每克纯的油或脂肪能提供9卡路里能量，等量淀粉则只能提供4卡路里能量，油脂供给的能量占每日能量摄入的10%~30%。膳食中即使没有添加油脂，食物中的天然脂肪也能提供10%~12%的总能量摄入。

油脂除了为身体能提供能量外还有其他功能。它们可以作为载体将脂溶性的维生素A、维生素D、维生素E和维生素K带入身体，并促进吸收。一些植物油含有人体不能合成的必需脂肪酸。脂肪能赋予食物特殊的风味和质地,从而增加适口性。脂肪还能增加就餐时的饱腹感,他们离开胃很慢,因此可能会延缓饥饿疼痛的复发。

油脂也被用作浅层烹调和深层油炸食物的介质。一些脂类化合物是不可或缺的食品乳化剂,另一些脂类化合物则可作为重要的油溶性颜料或食品着色剂。

脂质氧化

在食品的加工和保藏过程中,脂质会发生复杂的化学变化,并与其他食品成分发生反应,从而产生许多对食品质量有利和有害的化合物。脂质中不饱和脂肪酸的氧化是导致脂质降解的主要反应,也是食物腐败的主要因素之一。事实上,油脂的氧化水平是食品工业中的一个重要质量指标。油脂氧化不仅产生腐臭味,并且形成可能致病的氧化产物,降低了食品的安全性和营养质量。精制油必须储存在惰性气体如氮气或进行真空包装以防止氧化,并可以添加少量的防止脂肪氧化的化学物质——抗氧化剂。

Exercise

1. Answer questions

(1) What is the function of lipid?

(2) What is the structure of fat?

2. Translation

(1) Lipids in food exhibit unique physical and chemical properties. Their composition, crystalline structure, melting properties, and ability to associate with water and other non lipid molecules are especially important to their functional properties in many foods.

(2) Lipids consist of a broad group of compounds that are generally soluble in organic solvents but only sparingly soluble in water. They are major components of adipose tissue, and together with proteins and carbohydrates, they constitute the principal structural components of all living cells.

Lesson 4

Reading Material

Protein

Proteins play a central role in biological systems. They are vital to basic cellular and body functions, including cellular regeneration and repair, tissue maintenance and regulation, hormone and enzyme production, fluid balance, and the provision of energy. In addition, proteins are important constituents of food and one of major nutrients. They directly contribute to the flavor of food and are precursors for aroma compounds and colors formed during thermal or enzymatic reactions in production, processing and storage of food. Other food constituents, such as carbohydrates, also take part in such reactions. Proteins also contribute significantly to the physical properties of food through their ability to build or stabilize gels, foams, emulsions and fibrillar structures.

Composition and Structure

Proteins are very complex and many have been purified and characterized. Proteins vary in molecular mass, ranging from approximately 5000 to more than a million Daltons. They are composed of elements including carbon, hydrogen, oxygen, nitrogen and sulfur. α-amino acids are the basic structural units of proteins. These amino acids consist of an α-carbon atom covalently attached to a hydrogen atom, an amino group ($-NH_2$), a carboxyl group ($-COOH$), and a side-chain R group as indicated below:

$$H-\underset{R}{\overset{COOH}{\underset{|}{\overset{|}{C_\alpha}}}}-NH_2$$

The R may be a hydrogen atom or a more complex group, giving rise to variety of about 23 or more amino acids present in plant and animal proteins. Some of these amino acids cannot be synthesized in the body, but are essential for maintenance in human beings. These amino acids that have to be supplied in the food are called essential amino acids. The remaining amino acids, which can be synthesized from others in the body, are termed as non essential amino acids, because our body does not have to depend for their supply on the foods we eat.

Animal proteins have a more balanced amino acid profile and contain notable amounts of both essential and nonessential amino acids. The fish proteins contain abundant essential amino acids. Plant proteins generally have lower content of some essential amino acids such as lysine and methionine. Some nonessential amino acids such as glutamic acid and aspartic acid are abundantly found in all proteins.

The differences in structure and function of these thousands of proteins arise from the sequence in which the amino acids are linked together via amide bonds. Literally, billions of proteins with unique properties can be synthesized by changing the amino acid sequence, the type and ratio of amino acids, and the chain length of polypeptides.

Proteins can be classified by their composition, structure, biological function, or solubility properties. For example, simple proteins contain only amino acids upon hydrolysis e. g. albumin, globulin, prolamin, and glutelin, but conjugated proteins also contain non-amino-acid components e. g. nucleoprotein and lipoprotein.

Source of Proteins

Plants are the primary source of proteins as they synthesise protein by combining nitrogen and water from soil with the carbon dioxide from air. Most animals and fish depend on plants to provide them proteins.

For practical purposes, food proteins may be defined as those that are easily digestible, nontoxic, nutritionally adequate, functionally usable in food products, and available in abundance. Traditionally, milk, meats (including fish and poultry), eggs, cereals, legumes, and oilseeds have been the major sources of food proteins. In China and some other Asian countries, tofu is the largest source of food protein. It is consumed fresh or dried, or fried in fat and seasoned with soy sauce. Wheat proteins, another source of food protein, containing albumins, globulins, gliadins, and glutenins, these four basic proteins depending on their varied solubility in different solvents.

Food sources of proteins are presented in Table 4.1.

Table 4.1　　　　　　　　　　Food Sources of Proteins

Food	Protein content/(g/100g)
Milk	3~4
Egg	13
Fish, Meat	15~26
Vegetable	4~8
Cereals	6~13

However, because of the burgeoning world population, nontraditional sources of proteins for human nutrition need to be developed to meet the future demand. The suitability of such new protein sources for use in foods, however, depends on their cost and their ability to fulfill the normal role of protein ingredients in processed and home-cooked foods. The functional properties of proteins in foods are related to their structural and other physicochemical characteristics. A fundamental understanding of the physical, chemical, nutritional, and functional properties of proteins and the changes these properties undergo during processing is essential if the performance of proteins in foods is to be improved, and if new or less costly sources of proteins are to compete with traditional food proteins.

Protein properties

Food preferences by human beings are based primarily on sensory attributes such as texture, flavor, color, and appearance. The sensory attributes of a food are the net effect of complex interactions among various minor and major components of the food. Proteins generally have a great influence on the sensory attributes of foods. For example, the sensory properties of bakery products are related to the viscoelastic and dough-forming properties of wheat gluten; the textural and succulence characteristics of meat products are largely dependent on muscle proteins (actin, myosin, actomyosin, and several water-soluble meat proteins); the textural and curd-forming properties of dairy products are due to the unique colloidal structure of casein micelles; and the structure of some cakes and the whipping properties of some dessert products depend on the properties of egg-white proteins.

On an empirical level, the various functional properties of proteins can be viewed as manifestations of two molecular aspects of proteins: hydrodynamic properties and protein surface - related properties. The functional properties such as viscosity (thickening), gelation, and texturization are related to the hydrodynamic properties of proteins, which depend on size, shape, and molecular flexibility. Functional properties such as wettability, dispersibility, solubility, foaming, emulsification, and oil flavor binding are related to the chemical and topo-graphical properties of the protein surface.

However, when proteins are exposed to heat, light or change in pH and other processing conditions, structural changes occur. These changes in structure of proteins are known as denaturation. Denaturation leads to change in solubility of proteins. It may be reversible, if conditions which cause it are mild, but mostly the changes which occur are irreversible. Often denaturation has a negative connotation, because it indicates loss of some properties. For example, many biologically active proteins lose their activity upon denaturation. In some instances, however, protein denaturation is desirable. For example, partially denatured proteins are more digestible and have better foaming and emulsifying properties than do native proteins. Thermal denaturation is also a prerequisite for heat-induced gelation of food proteins.

Vocabulary

actin　肌动蛋白
actomyosin　肌动球蛋白
amino acid　氨基酸
aroma　芳香,香味
burgeoning　迅速增长的
casein　酪蛋白
cellular　细胞的
curd　凝乳
dairy　乳制品
denaturation　变性
emulsion　乳液,乳剂
gel　凝胶,胶化
micelle　胶束,胶囊

myosin　肌球蛋白
nucleoprotein　核蛋白
lipoprotein　脂蛋白
polypeptide　多肽
precursor　前体
protein　蛋白质
regeneration　再生
repair　修复
solubility　溶解度
textural　质构化,纹理性
tissue　组织
viscoelastic　黏弹性
whipping　搅打…变稠

参考译文

蛋　白　质

蛋白质在生物系统中起核心作用,它对细胞的组成和身体功能至关重要,包括细胞再生和修复、组织维持和调节、激素和酶的产生、体液平衡和能量供应。此外,蛋白质是食物的重要组成部分,也是主要的营养物质之一,它有助于食品风味形成,是食品生产、加工和储存过程中热或酶反应时生成芳香物质和色素的前体物质,食品中的另一种组成成分——碳水化合物也参与了这种反应。蛋白质还可以通过构建稳定的凝胶、泡沫、乳液和纤维状结构,使食品具有一定物理特性。

组成和结构

蛋白质是非常复杂的,许多已被纯化和表征,其分子质量不等,一般在 5000 甚至超过 100 万 u,它的组成元素一般有碳、氢、氧、氮和硫。α-氨基酸是蛋白质的基本结构单位,它由一个 α-碳原子通过共价键与一个氢原子、一个氨基(—NH_2)、一个羧基(—COOH)和侧链 R 基组成,如下所示:

$$\begin{array}{c} \text{COOH} \\ | \\ \text{H}—\text{C}_\alpha—\text{NH}_2 \\ | \\ R \end{array}$$

R 基可以是氢原子或更复杂的基团,这样导致存在于植物和动物蛋白质中约 23 个或更多氨基酸的多样性。这里面有些氨基酸不能在人体内合成,但对维持人体生命活动至关重要,这些氨基酸必须从食品中获得,称之为必需氨基酸。另外一些氨基酸,可以在体内通过其他物质合成,我们的身体不必依赖食物的供给,这类氨基酸称为非必需氨基酸。

动物蛋白质含有大量的必需氨基酸和非必需氨基酸,具有更平衡的氨基酸谱。鱼类蛋白质中含有丰富的必需氨基酸。植物蛋白质所含必需氨基酸的含量通常较低,如赖氨酸和蛋氨酸。在所有蛋白质中,都含一些大量的非必需氨基酸如谷氨酸和天冬氨酸。

成千上万的蛋白质在结构和功能上的差异源于通过酰胺键连接的氨基酸序列的不同。从这方面看,通过改变氨基酸序列、氨基酸的类型和比例以及多肽的链长,可以合成出几十亿种具有特殊性质的蛋白质。

蛋白质可以根据其组成、结构、生物学功能或溶解特性进行分类。例如,简单蛋白质水解后仅含有氨基酸,如清蛋白、球蛋白、醇溶蛋白和谷蛋白;但结合蛋白质水解后还含有非氨基酸组分,如核蛋白和脂蛋白。

蛋白质来源

植物是蛋白质的主要来源,植物通过利用来自土壤中氮和水以及空气中的二氧化碳来合成蛋白质。大部分动物和鱼类则依赖植物来满足其蛋白质需求。

根据实际需求,食物蛋白质被定义为易消化、无毒、营养充分、具备功能性的食物中的蛋白质,并且容易获得。传统意义上,主要的食物蛋白质来源有奶类、肉类(包括鱼和家禽)、蛋类、谷类、豆类和油籽。在中国和其他一些亚洲国家,豆腐是食物蛋白的最大来源,新鲜、干制或油炸后调味均可食用。小麦蛋白也是一种食物蛋白,根据溶解性可分为白蛋白、球蛋白、醇溶蛋白和麦谷蛋白。

常见食物蛋白的来源见表 4.1。

表 4.1　　蛋白质的常见食物来源

食物	蛋白质含量/(g/100g)
牛乳	3~4
鸡蛋	13
鱼,肉类	15~26
蔬菜	4~8
谷物	6~13

然而,由于世界人口的不断增长,需要开发非传统的蛋白源食物来满足未来人们对营养的需求。不过,这种新的蛋白源食物的适用性还取决于它们的成本,以及在加工和烹饪后满足正常蛋白质成分的功能。食物中蛋白质的功能与其结构和理化性质有关。如果要改善食物中蛋白质的功能,并且,这种新的或低成本的蛋白源食物能与传统蛋白食物竞争,那么,对蛋白质的物理、化学、营养和功能特性以及这些特性在加工过程中发生的变化有一个基本的了解是非常必须的。

蛋白质性质

人们对食品的偏好主要是基于其感官特性,如质地、风味、颜色和外观。食品的感官特性是食物中各种次要和主要成分之间复杂相互作用后的净效应。通常蛋白质对食品的感官特性有很大的影响。例如,烘焙食品的感官特性与面筋蛋白的黏弹性和成团性有关;肉制品的纹理性和多汁性在很大程度上依赖于肌蛋白(肌动蛋白、肌球蛋白、肌动球蛋白和多种水溶性肉蛋白);乳制品的质构特性和凝乳性与酪蛋白胶束独特的胶体结构有关;一些糕点的结构和一些甜点的搅打特性取决于蛋清蛋白的性质。

从经验上看,蛋白质的各种功能性质可以看作是两种蛋白质分子特性的体现:流体动力学性质和蛋白质表面相关性质。蛋白质的一些功能性质如黏性(增稠)、凝胶化和组织化与蛋白质的流体动力学性质有关,这取决于蛋白质大小、形状和分子的灵活性。另一些功能性质如润湿性、分散性、溶解性、发泡性、乳化性以及和油脂风味的结合与蛋白质表面的化学和外表特性有关。

然而,当蛋白质所处的热、光、pH、工艺条件发生变化时,会发生结构变化,蛋白质这种结构变化称为变性。变性会导致蛋白质溶解度的变化,如果条件温和,这种变性是可逆的,但大多数变性是不可逆的。变性常常具有负面意义,因为这意味着一些性质的丧失。例如变性会导致很多蛋白质的生物活性丧失。不过,有些蛋白质的变性是必须的。例如,部分蛋白质变性后比天然蛋白质更易消化,具有更好的起泡性和乳化性。热变性也是食物蛋白通过热诱导凝胶化的先决条件。

Exercise

1. Answer questions

(1) What are proteins? What are their functions?

(2) List the different sources of protein in our dietary.

(3) List the different milk products used in our diets.

2. Translation

(1) Proteins are important constituents of food and one of major nutrients. They directly contribute to the flavor of food and are precursors for aroma compounds and colors formed during thermal or enzymatic reactions in production, processing and storage of food.

(2) Plants are the primary source of proteins as they synthesise protein by combining nitrogen and water from soil with the carbon dioxide from air. Most animals and fish depend on plants to provide them proteins.

(3) When proteins are exposed to heat, light or change in pH and other processing conditions, structural changes occur. These changes in structure of proteins are known as denaturation.

Lesson 5

Reading Material

Vitamin

Vitamins are organic compounds present in small amounts in foods, which must be provided to the body, to ensure normal growth and maintenance of the body. Now we know that vitamins are one of the six classes of nutrients supplied by food. Vitamins are important for the body regulatory and protective functions. Unlike most other nutrients, they are required in very small amounts. With few exceptions, humans cannot synthesize most vitamins and therefore need to obtain them from food and supplements. Insufficient levels of vitamins result in deficiency diseases (e. g., scurvy and pellagra, which are due to the lack of ascorbic acid (vitamin C) and niacin, respectively).

The main exceptions are vitamin D, a hormone-like vitamin that can be synthesized in the skin upon exposure to sunlight and other forms of ultraviolet radiation, and therefore is not absolutely required in the diet, and vitamin K, which can be synthesized by intestinal microflora. Some vitamins have multiple chemical forms or isomers that vary in biological activity. For example, four forms of tocopherol occur in food, with α-tocopherol having the greatest vitamin E activity. Numerous carotenoids occur in nature and in food, but only a few have vitamin A activity. Some carotenoids, such as lycopene, do not possess vitamin A activity but, nonetheless, have important health benefts.

Classification

Vitamins are conveniently classified into two groups on the basis of their solubility (in fat or in water) into fat-soluble and water-soluble vitamins. Fat-soluble vitamins include A, D, E and K, and water-soluble vitamins include the B-group [e.g. vitamins B_1(thiamin), B_2(riboflavin), B_3(niacin), B_5(pantothenic acid), B_6(pyridoxine), B_{12}(cyanocobalamin), biotin, and folic acid] and vitamin C. Foods differ greatly in the amount and kinds of the vitamins they supply.

Fat-soluble vitamins are capable of being stored in the body and can accumulate. They are not easily destroyed by heat during cooking or processing or through exposure to air. In contrast, water-soluble vitamins usually do not accumulate in the body because they are stored only in small quantities. However, this property allows deficiencies of water-soluble vitamins to occur faster than deficiencies of fat-soluble vitamins. Excessive dosages of water-soluble vitamins generally are excreted in the urine but can lead to toxicity if the dosages are high enough.

Function

The vitamins comprise a diverse group of organic compounds that are nutritionally essential micronutrients. Vitamins function in vivo in several ways, including (a) as coenzymes or precursors (niacin, thiamin, riboflavin, biotin, pantothenic acid, vitamin B_6, vitamin B_{12}, and folate); (b) as components of the antioxidative defense system (ascorbic acid, certain carotenoids, and vitamin E); (c) as factors involved in genetic regulation (vitamins A, D, and potentially several others); and (d) with specialized functions such as vitamin A in vision, ascorbate in various hydroxylation reactions, and vitamin K in specific carboxylation reactions.

Most of the vitamins exist as groups of structurally related compounds exhibiting similar nutritional function. Various forms of each vitamin can exhibit vastly different stability (e.g., pH of optimum stability and susceptibility to oxidation) and reactivity. For example, tetrahydrofolic acid and folic acid are two folates that exhibit nearly identical nutritional properties.

Toxicity

In addition to the nutritional role of vitamins, it is important to recognize their potential toxicity. Vitamins A, D, and B_6 are of particular concern in this respect. Episodes of vitamin toxicity are nearly always associated with overzealous consumption of nutritional supplements. Potential toxicity also exists from inadvertent excessive fortification, as having occurred in an incident with vitamin D-fortified milk. Instances of intoxication from vitamins occur endogenously in food are exceedingly rare.

Sources of Vitamins

Although vitamins are consumed in the form of supplements by a growing fraction of the population, the food supply generally represents the major and most critically important source of vitamin intake. Foods, in their widely disparate forms, provide vitamins that occur naturally in plant, animal, and microbial sources as well as those added in fortifications. In addition, certain dietetic and medical foods, enteric formulas, and intravenous solutions are formulated so that the entire vitamin requirements of the individual are supplied from these sources.

The popular literature has emphasized that natural vitamins are superior to synthetic ones, even though the active ingredient is often chemically identical in both forms. However, added folic acid in dietary supplements or fortified foods is better absorbed than naturally occurring folates in foods. Pregnant women administered synthetic folic acid (pteroylglutamic acid) had higher serum folate levels than women given the natural conjugated form of the vitamin.

Regardless of whether the vitamins are naturally occurring or added, the potential exist for losses by chemical or physical (leaching or other separations) means. Losses of vitamins are, to some degree, inevitable in the manufacturing, distribution, marketing, home storage, and preparation of processed foods, and losses of vitamins, can also occur during the post-harvest handling and distribution of fruits and vegetables and during the post-slaughter handling and distribution of meat products. Since the modern food supply is increasingly dependent on processed and industrially formulated foods, the nutritional adequacy of the food supply depends, in large measure, on our understanding of how vitamins are lost and on our ability to control these losses.

Vocabulary

ascorbic acid　抗坏血酸(维生素 C)　　　　fat-soluble　脂溶性
carotenoids　类胡萝卜素　　　　　　　　　intestinal　肠道的

microflora 微生物群落
niacin 烟酸
regulatory 调节的
supplement 补充剂
tocopherol 生育酚
thiamin 硫胺素(维生素 B_1)
vitamin 维生素

参考译文

维 生 素

维生素是食品中含量较低的一类有机化合物,它们是机体维持正常生理功能所必需的。我们知道维生素是食物中的六大营养物质之一,它们对机体有重要的调节和保护功能。与其他大多数营养物质不同,维生素的需求量非常少。除个别维生素外,人类不能合成大多数维生素,需要从食物和补充剂中获得。维生素缺乏会导致一些疾病[如坏血病和糙皮病分别是由于抗坏血酸(维生素 C)和烟酸的缺乏]。

维生素 D 是个例外,它是一种类似激素的维生素,皮肤在阳光和其他形式的紫外线照射下能合成维生素 D,因此饮食中并不是绝对需要维生素 D,维生素 K 则可以通过肠道微生物合成。有些维生素有多种化学形式或异构体,生物活性各异。例如,生育酚在食品中有四种形式,其中 α-生育酚具有最大的维生素 E 活性。许多类胡萝卜素存在于自然界和食物中,但只有少数有维生素 A 活性。一些类胡萝卜素如番茄红素等,虽不具有维生素 A 的活性,但对健康大有益处。

分类

根据维生素溶解性(溶于脂肪或水)可将其简单的分成脂溶性和水溶性维生素两类。脂溶性维生素包括维生素 A、维生素 D、维生素 E、维生素 K,水溶性维生素包括 B 族维生素和维生素 C,其中 B 族维生素如维生素 B_1(硫胺素)、维生素 B_2(核黄素)、维生素 B_3(烟酸)、维生素 B_5(泛酸)、维生素 B_6(吡哆醇)、维生素 B_{12}(氰钴胺素)、生物素和叶酸。不同食物所含维生素的含量和种类不同。

脂溶性维生素能够储存在生物体内,因此可以积累。它们在烹饪、加工或暴露于空气时不易被热量破坏。相比之下,水溶性维生素通常不会积聚在体内,因为它们只能少量储存。然而,这种性质使水溶性维生素的缺乏症比脂溶性维生素的缺乏症的发生率更大。过量的水溶性维生素通常在尿中排出,但如果剂量足够高,则会引起毒性。

功能

维生素是由各种各样的有机化合物构成的,它们是营养必需的微量营养物质。维生素在体内的功能有:(a)作为辅酶或前体物质(如烟酸、硫胺素、核黄素、生物素、泛酸、维生素 B_6,维生素 B_{12},叶酸);(b)作为抗氧化防御系统的组成部分(抗坏血酸、类胡萝卜素、维生素 E);(c)作为基因调控因子(维生素 A、维生素 D,以及潜在的其他维生素);(d)一些特殊功能,如维生素 A 与视觉,抗坏血酸的各种羟基化反应,维生素 K 的特殊羧基化反应。

大多数维生素以一组结构相似的化合物形式存在,它们具有相似的营养功能。每种维生素的不同形式则表现出不同的稳定性(如最稳定 pH,氧化敏感性)和反应活性。例如,四氢叶酸和叶酸具有几乎相同的营养特性。

毒性

维生素除了具有营养功能外,认识到它们的潜在毒性也是非常重要的。在这方面,维生素 A、维生素 D 和维生素 B_6 需要特别注意。维生素中毒几乎总是与过量摄取营养补充剂有关。维生素体现潜在毒性也存在于无意间的过度强化,如维生素 D 强化牛乳的中毒事件就发生过。在食物中因内源性维生素中毒的案例极少发生。

维生素来源

随着人口增长,维生素常以补充剂形式来补充摄入,但通过食物供应是主要的和最重要的维生素来源。食物中维生素的来源形式广泛且多样,包括从植物、动物和微生物中获取,以及那些添加了维生素的强化剂。此外,也出现了一些饮食和医疗食品、肠溶配方和静脉注射液的维生素补充源,来满足个体对全部维生素的需求。

比较流行的说法是天然维生素优于合成的维生素,即使它们的活性成分的化学性质相同。然而,叶酸的膳食补充剂或强化食品比食品中的天然叶酸更易吸收。孕妇服用合成叶酸(叶酸)后,血清中的叶酸水平高于服用天然共轭形式的叶酸维生素。

无论维生素是自然存在还是添加的,化学或物理(浸出或其他分离方法)方式都有可能造成维生素损失。一定程度上,在食品的生产、分销、营销、家庭储藏和加工食品的准备过程中维生素的损失都是不可避免,并且维生素的损失也可能发生在果蔬的采后处理、分装及肉制品屠宰后的处理和流通过程中。由于现在的食品供应越来越依赖加工和工业配制的食品,这些食品的营养充足程度在很大程度上取决于维生素的损失程度以及我们控制这些损失的能力。

Exercise

1. Answer questions

(1) What is Vitamin? What are the common features of vitamins?

(2) What are the functions of vitamin?

(3) List three important dietary sources of the following vitamins:

(a) Vitamin A

(b) Thiamin

(c) Riboflavin

(d) Niacin

(e) Ascorbic acid

2. Translation

(1) Vitamins are organic compounds present in small amounts in foods, which must be provided to the body, to ensure normal growth and maintenance of the body.

(2) Vitamins are conveniently classified into two groups on the basis of their solubility (in fat or in water) into fat-soluble and water-soluble vitamins. Fat-soluble vitamins include A, D, E and K, and water-soluble vitamins include the B-group [e.g. vitamins B_1(thiamin), B_2(riboflavin), B_3(niacin), B_5(pantothenic acid), B_6(pyridoxine), B_{12}(cyanocobalamin), biotin, and folic acid] and vitamin C.

Lesson 6

Reading Material

Mineral

In general, minerals are the constituents which remain as ash after the combustion of plant and animal tissues. Human body uses minerals to activate the enzymes, hormones, and other molecules that participate in the function and maintenance of life processes. Mineral elements are present in relatively low concentrations in foods. Nevertheless, they play key functional roles in both living systems and foods. Minerals are absolutely essential, as they cannot be synthesized by the body.

Chemical properties of minerals

Many different chemical forms of mineral elements are present in foods. These forms are commonly referred to as species and include compounds, complexes, and free ions. Given the diversity of chemical properties among the mineral elements, the number and diversity of nonmineral compounds in foods that can bind mineral elements, and the chemical changes that occur in foods during processing and storage, it is not surprising that the number of different mineral species in foods is large indeed. Moreover, since foods are so complex and many mineral species are unstable, it is very difficult to isolate and characterize mineral species in foods. Thus, our understanding of the exact chemical forms of minerals in foods remains limited. Fortunately, principles and concepts from the vast literature in inorganic, organic, and biochemistry can be very useful in guiding predictions about the behavior of mineral elements in foods.

Function and classfication

Minerals have several functions in the body. Body building minerals serve as structural constituents in the hard tissues of the body, such as the bones and teeth. Thus, the minerals are needed in comparatively large amounts to help normal growth and development of bones and teeth. Minerals are also the components of soft tissues (muscle, nervous tissue). They form a part of compounds essential for body functions (haemoglobin, thyroxine, insulin etc.)

Some minerals help to maintain acid-base balance and while others regulate water balance in the body. Some minerals are important in transmission of nerve impulses, while others are necessary for muscle contraction and relaxation. Some metals are integral part of enzymes, while others function as cofactors of some enzymes.

The number of different minerals necessary to maintain good health is still unknown. While dietary deficiencies of only a few minerals such as iron or iodine can be linked to disease directly, some authorities claim that as many as 60 elements are necessary for optimum longevity and quality of life in humans. What has become more widely recognized is that a number of minerals are required by the body, and deficiencies in these minerals may not produce obvious symptoms but can still result in poor health or a shortened life expectancy. Even some minerals once considered solely "toxic" have now been identified as important in supporting longevity and quality of life. Selenium was long viewed as a toxic compound, but several international studies have demonstrated that dietary selenium markedly reduces the incidences and death of cancer. Some forms of chromium are also considered toxic, but trivalent chromium is a mineral associated with fat metabolism and has been linked to the regulation of blood glucose levels in the body.

There are two groups of minerals, major minerals and trace minerals. Major minerals (also known as macrominerals or macroelements) are needed in the diet in amounts of 100mg or more each day. They include calcium, magnesium, sodium, potassium and phosphorus. Macrominerals are present in virtually all cells of the body, maintaining general homeostasis, and are required for normal functioning. Acute imbalances of these minerals can be potentially fatal, although nutrition is rarely the cause of these cases. Trace minerals (also known as microminerals) are micronutrients that are chemical elements. They include iron, chromium, copper, iodine, manganese, selenium, zinc, and molybdenum. They are dietary minerals needed by the human body in very small quantities (generally less than 100 mg/day).

There are several other minerals that may be essential for humans, but research has not established their importance, including tin, nickel, silicon and vanadium. There are also minerals found in the body that are regarded as contaminants including lead, mercury, arsenic, aluminum, silver, cadmium, barium, strontium and others.

Solubility of Minerals in Aqueous Systems

All biological systems contain water, and most nutrients are delivered to and metabolized by organisms in an aqueous environment. Thus the availabilities and reactivities of minerals depend, in large part, on solubility in water. This excludes the elemental form of nearly all elements from physiological activity in living systems since these forms, such as elemental iron, are insoluble in water and therefore are unavailable for incorporation into organisms or biological molecules.

The solubilities of mineral complexes and chelates may be very different from that of inorganic salts. For example, if ferric chloride is dissolved in water, the iron will soon precipitate as ferric hydroxide. On the other hand, ferric iron chelated with citrate is quite soluble. Conversely, calcium as calcium chloride is quite soluble, while calcium chelated with oxalate ion is insoluble.

Minerals in Food Processing

Generally, there are metal ions, derived from food itself or acquired during food processing and storage, which interfere with the quality and visual appearance of food. They can cause discoloration of fruit and vegetable products and many metal-catalyzed reactions are responsible for losses of some essential nutrients, for example, ascorbic acid oxidation. Also, they are responsible for taste defects or off-flavors, for example, as a consequence of fat oxidation. Therefore, the removal of many interfering metal ions by chelating agents or by other means is of importance in food processing.

Vocabulary

calcium　钙
chelate　螯合
cofactor　辅因子
copper　铜
chromium　铬
discoloration　变色
homeostasis　动态平衡

hormone　激素
inorganic　无机的
iodine　碘
iron　铁
magnesium　镁
manganese　锰
mineral　矿物质

molybdenum	钼	silicon	硅
potassium	钾	sodium	钠
phosphorus	磷	symptoms	症状
selenium	硒	zinc	锌

参考译文

矿 物 质

矿物质通常指动植物组织经过燃烧后剩余的灰分。人体能利用矿物质激活参与维持机体生理功能的酶、激素和其他分子。食物中的矿物质浓度也相对较低，尽管如此，它们在生命系统和食品中都起着重要作用。由于人体自身不能合成矿物质，因此它们是人体必不可少的元素。

矿物质化学性质

矿物质元素以不同的形式存在于食品中，包括化合物、复合物和游离离子的形式。鉴于矿物质元素的化学性质多样，以及食品中能够结合矿物质元素的非矿物质化合物的数量和种类都多，并且食品在加工和储存过程中也会发生化学变化，因此，食品中拥有如此多矿物质形态就不足为奇了。此外，由于食品成分复杂，许多矿物质形态也不稳定，因此很难分离和鉴定食品中的矿物质。因此，我们对食品中矿物质的准确化学形态的了解仍然有限。幸运的是，无机、有机和生物化学的大量文献中的原理和概念对于指导和预测食品中矿物质元素的形态非常有用。

功能和分类

矿物质对机体有多种功能，它们是构成机体结构组织的成分，如骨骼和牙齿，因此，机体需要大量的矿物质来帮助骨骼和牙齿的正常生长发育。矿物质也是软组织的组成成分(如肌肉、神经组织)，它们是维持机体功能必需化合物的组成成分(如血红蛋白、甲状腺素、胰岛素等)。

有些矿物质有助于维持机体酸碱平衡，其他一些矿物质则可调节机体内的水分平衡。有些矿物质对神经兴奋的传导有重要作用，而另一些矿物质则是肌肉收缩和放松所必需的物质。还有些金属元素是酶的组成成分，有的还可以作为酶的辅助因子。

维持身体健康所必需的矿物质数量还是未知的。虽然膳食中只有少数矿物质如铁或碘的缺乏直接与一些疾病相关，但权威人士认为，人类要想长寿和拥有好质量的生活则需要多达 60 种的矿物质元素。众所周知,人体需要多种矿物质,这些矿物质的缺乏可能不会产生明显的症状,但仍然会导致健康状况不佳或者预期寿命缩短。即使一些曾被认为是"有毒"的矿物质,现在已经确定它们对长寿和生活质量的影响非常重要。硒就被长期视为一种有毒物质,但一些国际研究表明,膳食中的硒显著降低了癌症的发生率和死亡率。某些形式的铬也被认为是有毒的,但三价铬与脂肪的代谢有关,并且与体内血糖水平的调节有关。

矿物质可分为主要矿物质和微量矿物质。主要矿物质也称为大量矿物质或常量元素,每日膳食中大量矿物质需要量为 100mg 以上。它们包括钙、镁、钠、钾、磷。大量矿物质几乎存在于身体所有细胞中,用来维持体内动态平衡和正常功能。这些矿物质的急性失衡可能是致命的,虽然营养上它们不是导致这些病例的原因。微量矿物质(也称微量元素)是化学元素中的微量营养物质,包括铁、铬、铜、碘、锰、硒、锌和钼。它们是人们膳食中必需的矿物质,需求量非常少(通常小于 100mg/d)。

还有其他几种可能对人类至关重要但研究尚未确定其重要性的矿物质,包括锡、镍、硅和钒。人体中还存在些被认为污染物的矿物质,如铅、汞、砷、铝、银、镉、钡、锶等。

矿物质在水中的溶解度

所有生物系统都含有水分,生物体中的大多数营养物质都是在水环境中进行传递和代谢的。因此,矿物质的可利用性和反应活性在很大程度上取决于其在水中的溶解度。但这不包括生物系统中有生理活性的矿物质元素的基本形态,因为这些形态如元素铁不溶于水,它不能融入生物体或生物分子。

矿物质的复合物和螯合物的溶解度可能与无机盐非常不同。例如,如果氯化铁溶解在水中,铁很快会以氢氧化铁形式沉淀,而另一方面,三价铁与柠檬酸螯合后则易溶于水。相反,钙作为氯化钙时易溶于水,而与草酸根离子螯合的钙则不溶。

矿物质与食品加工

通常,食物自身含有的或加工和储存过程中获得的金属离子会影响食品的质量和外观。它们可能引起水果和蔬菜产品的变色,并且许多金属催化的反应会导致必需营养物质的损失,如抗坏血酸氧化。此外,它们也会导致味觉缺陷或异味,如脂肪氧化后。因此,食品加工中利用螯合剂或其他方法去除许多干扰的金属离子具有重要意义。

Exercise

1. Answer questions
(1) What are the functions of minerals?
(2) Which are the major minerals and which are the trace minerals?
2. Translation
(1) In general, minerals are the constituents which remain as ash after the combustion of plant and animal tissues. Human body uses minerals to activate the enzymes, hormones, and other molecules that participate in the function and maintenance of life processes.

(2) Minerals have several functions in the body. Body building minerals serve as structural constituents in the hard tissues of the body. Some minerals help to maintain acid-base balance and while others regulate water balance in the body. Some minerals are important in transmission of nerve impulses, while others are necessary for muscle contraction and relaxation. Some metals are integral part of enzymes, while others function as cofactors of some enzymes.

Unit II
Food Processing
食品加工

Lesson 7

Reading Material

Bread Making

Bread is one of the oldest prepared foods. It is not only consumed in all countries of the world, but also is the most acceptable form of food there. Bread is a staple food prepared from a dough of flour and water, usually by baking. There are many kind of bread peculiar to the respective countries. For example, there are French, German, British and other bread, each of which has its own characteristic in materials, method of making and tastes. Bread is bond to become more popular throughout the word.

Raw Material for Bread

(1) Wheat Flour

Wheat Flour provides bulk and structure to the bread. Wheat flour, the main raw material for bread, is broadly classified into high protein (protein over 11%), medium protein flour (protein 9% ~ 11%) and low protein (protein below 9%).

Flour contains two proteins, gliadin and glutenin, which can combine and produce gluten in the presence of water. Gluten is elastic, so it helps to hold the air and carbon dioxide in the mixture and make it light. In bread making, the gluten is developed by kneading to form a strong network which sets on baking. Different wheat has varying amounts of gluten. Looking for them labeled as bread flour or even high gluten flour. Whole meal flour can be used for bread making but bread will be coarse in the texture dark in color.

(2) Yeast

Yeast is a leavening agent that produces carbon dioxide, which makes the bread rise. Yeast are single celled organisms that live solely to eat sugars and reproduce. When mixed with water and sugar, the yeast ferments to produce carbon dioxide, filling the bread dough with tiny air bubbles. You can choice fresh yeast, active dry yeast and instant, or rapid rise yeast. It is convenient to use dry yeast because its handing is simple and its preservation easy. Fresh yeast can get moldy if left too long in the refrigerator, and can be temperamental.

It is important to remember that:

① Approximately half as much dried yeast as fresh is required in any recipe.

② Reconstitution is necessary before using dried yeast. This is done by soaking the yeast in lukewarm water (29.4℃) containing 14.2g sugar to 0.28L water. The water and sugar used for reconstitution should be subtracted from the quantities given in the recipe.

(3) Water

Water, or some other liquid, is used to form the flour into a paste or dough. Water provides for gluten formation and yeast reproduction. For bread making, medium hard water works better. As for water, there is no problem generally, if city water is available. However, in the case of drawing water from well, hard water is not desirable. Water with 5~7 pH is best. Most bread does best when the ratio of flour to water is 2 : 1.

(4) Other materials

Salt: Salt regulates yeast growth and gives flavor. It also affects the crumb and the overall texture by stabilizing and strengthening the gluten. Without the presence of salt, yeast growth will continue until the flour matrix can no longer support it, and bread will deflate. Too much salt, yeast won't give bread enough rise. Too little, bread will rise too much. The trick is in getting the right salt to yeast ratio. The proportion of salt used also varies but it is common practice to use about 2.3g per

127.0g of flour.

Fats or shortenings: Fats, such as butter, vegetable oils, lard, or that contained in eggs, affect the development of gluten in breads by coating and lubricating the individual strands of protein. They also help to hold the structure together. If too much fat is included in a bread dough, the lubrication effect causes the protein structures to divide. A fat content of approximately 3% by weight is the concentration that produces the greatest leavening action. In addition to their effects on leavening, fats also serve to tenderize breads and preserve freshness.

Sugar: Acts as a yeast food and increases tenderness and browning, and keeping qualities.

Liquid: Instead of water, recipes may use liquids such as milk or other dairy products (including buttermilk or yogurt), fruit juice, or eggs. These contribute additional sweeteners, fats, or leavening components, as well as water. Milk doughs have a finer texture, a better flavor and brown more quickly. Milk doughs also help make a complete protein.

Bread improvers: Bread improvers and dough conditioners are often used in producing commercial breads to reduce the time needed for rising and to improve texture and volume. The substances used may be oxidising agents to strengthen the dough or reducing agents to weaken gluten and reduce mixing time, emulsifiers to strengthen the dough or to provide other properties such as making slicing easier, or enzymes to increase gas production.

As for the material for packaging, some countries are using polyethylene. But many countries use wax paper or grease proof paper.

Process steps

There are quite a few steps to making bread. Here they are: scaling, mixing, fermenting, dividing, rounding, baking and cooling.

Some plants to manufacture bread involve the following sequence of operation:

(1) Air sifter with pneumatic conveyor

Flour should flow into the air sifter to be sifted and to eliminate obstructions and is conveyed to the hopper of the mixer by the Pneumatic conveyor.

(2) Mixing

Dough is made from flour and such materials as yeast, sugar, and salt, jot, yeast food etc. After mixing, tear off a small piece of dough. The dough becomes thin enough that it gets nice and translucent when tearing it. This type of mixer has two steps of speed, low and high. For mass production, a high-speed horizontal mixer will be suitable. The dough is fermented in the room for 2 to 4 hours, and again mixed at is stage.

(3) Dough dividing

Mixed dough is weighed and cut by dough divider.

(4) Rounding

The weighed and cut dough pieces are carried by the conveyor belt of the dough divider and are rounded with the rounder.

(5) Middle proofing

Rounded dough pieces are conveyed by a bucket convey and to be fermented for about 15 minutes to relax gluten and make handling easier.

(6) Moulding

Fermented dough pieces are transferred by shooter and moulded by moulding machine.

(7) Second fermenting or proofing

Eased dough pieces on racks are carried into second fermentation room which has a temperature of 38~40℃ and humidity of 80%~85%. The proving period normally occupies from 25 to 40 minutes according to the type of bread being made. After second fermentation, dough surface are brushed with egg wash, water, milk or egg white.

(8) Baking

Fermented dough pieces are baked by a preheated oven. Baking temperature is 232.2 to 260℃, and the baking time from 40 to 50 minutes.

(9) Cooling

Baked products are conveyed by cooling conveyor and they are cooled during conveying.

(10) Slicing and wrapping

Cooled Products are sliced and wrapped automatically.

Process description

For best results, using high-protein bread flour is better. Flour too low in protein produces a loaf that is poor in volume and texture. If the flour is old, it will cause a crumbly, "short" dough. Flour should be stored in airtight containers in a cool, dry place (less than 60% humidity). All-purpose, bread and cake flour will keep for 6 months to a year at 21.1℃ and 2 years at 4.4℃. Whole-wheat flour should be refrigerated or frozen, if possible. Before using refrigerated or frozen flour, allow it to warm to room temperature. With bread making, exact flour measurements are impossible. Dough is affected by heat, humidity, sugar, and also affected by altitude, possibly the personality and the mood of the baker. If too much flour is used, the bread may be very heavy and stiff. If too little is used, the bread will not hold up and a low-volume bread will result.

The water need to make the dough brought to a temperature that will give the finished dough a temperature of 23.9 to 27.6℃. The exact temperature required depending upon the proposed length of the fermentation. So, on hot days, cooler water may be used; on colder days, warmer ones. The necessary water temperature can be arrived by subtracting the temperature of the flour from twice the desired dough temperature.

Salt should not be omitted because it controls the action of yeast. Besides having a very bland flavor, breads made without salt tend to over-rise and will have a different texture than breads with salt.

When adding wheat bran, wheat germ, bulgur or cracked wheat to a bread recipe, leave the bread dough as moist as possible, because these ingredients absorb liquid and tend to produce a drier loaf. Reduce the amount of kneading to avoid cutting the gluten strands with the sharp edges of these products. As the ratio of whole-wheat flour to bread flour increases, so does the rising time.

The time required for dough development varies considerably, depending on the factors such as temperature, humidity, yeast characteristics, flour characteristics and the kneading. To test if dough is sufficiently kneaded, poke the dough with your fingers; it should spring back. Sometimes blisters will form on the surface of the dough, which is another sign the dough is sufficiently kneaded.

Vocabulary

gliadin　麦胶蛋白
glutenin　麦谷蛋白
gluten　面筋
knead　揉,按摩,捏(面团、湿黏土等),揉捏(肌肉等)
temperamental　气质的,性格的,不可靠的
reconstitution　重新组建,重新构建
lukewarm　(液体)微温的
utensil　器具,用具,器皿

crumb　面包心,面包屑
matrix　基质
shortening　起酥油
lubricating　涂油
conditioners　调节剂
polyethylene　聚乙烯
pneumatic conveyor　气动输送机
hopper　加料斗,送料斗,漏斗
dough　(用于制面包和糕点的)生面团

high-speed horizontal mixer　高速卧式搅拌机　　bulgur　碾碎的干小麦
rack　支架,架子　　proving period　醒发阶段
loaf　一条(块)面包

参考译文

面包制作

面包是最古老的食品之一。它不仅被世界各国人民所食用,而且是一种易被接受的食物。面包是由面粉和水制成面团后,经焙烤而成的一种主食。各个国家都有许多特色面包,如法式、德式、英式以及一些其他类面包,每种面包在原材料、制作方法和口味方面都有自己的特点。面包在全世界广受欢迎。

面包原料

(1)小麦面粉

小麦面粉构成面包的主体和结构,是制作面包的主要原料,它大致可分为高蛋白质(蛋白质含量高于11%)、中等蛋白质(蛋白质9%~11%)和低蛋白质(蛋白质低于9%)。面粉含有麦胶蛋白和麦谷蛋白两种蛋白质,在有水分存在时,它们结合并形成面筋。面筋有弹性,从而有助于空气和二氧化碳保持在混合料中,使混合料变蓬松。在面包制作过程中,通过揉捏形成面筋,由面筋形成一定强度的网络,在烘烤时固定下来。不同的小麦含有不同量的面筋,寻找标有面包粉或高筋粉的面粉。全麦面粉可以用来制作面包,但全麦粉制的面包质地粗糙、色暗。

(2)酵母

酵母是一种能产生二氧化碳的膨松剂,使面包的体积增大。酵母是单细胞微生物,仅靠糖就能存活和繁殖。当水和糖混合时,酵母发酵产生二氧化碳,小气泡充满面包面团。可以选择鲜酵母、干酵母、即发活性干酵母或快速发酵酵母。干酵母使用方便,操作简单,易储藏,鲜酵母长时间储藏在冰箱里会发霉,并且产生霉味。

以下两点值得记住:

①在任何面包配方中,鲜酵母的量大约是干酵母的一半。

②干酵母在使用前必须活化。做法是:将酵母浸泡在含有14.2g糖的0.28L温水(29.4℃)中,活化用的糖和水,其量应当从配方中所给的量中扣除。

(3)水

水或其他液体可使面粉制成面团。水利于面筋的形成和酵母的繁殖。制作

面包用中等硬度水较好。至于水,如果用自来水,一般不成问题。但是,若使用地下井水,其硬度则不适合。pH 5~7 的水最合适。当面粉和水的比例为 2∶1 时,大部分面包做的最好。

(4)其他材料

盐:调节酵母的生长并赋予风味。它还通过稳定和强化面筋进而对面包屑和整个质地产生影响。如果没有盐,酵母将会不断生长,直到面粉基质不再支持它,这时面包会放气而缩小。盐太多,酵母不足以支撑面包膨胀,盐太少,面包会膨胀太多。诀窍就是要有合适的盐和酵母比例。各种面包所用盐的比例各不相同,通常的做法是每 127.0g 面粉用 2.3g 盐。

脂肪或起酥油:脂肪,如黄油、植物油、猪油或鸡蛋中含有的脂肪,通过包衣和润滑单个蛋白质链,影响面包中面筋的形成。它们也有助于保持结构。如果面包面团中含有过多的脂肪,润滑作用会导致蛋白质结构分裂。约3%质量的脂肪含量会产生最大的发酵作用浓度。除了对发酵的影响,脂肪也起到嫩化面包和保鲜的作用。

糖:作为酵母的食物,增加嫩度和褐变,保持品质。

液体:配方中可以使用牛乳或其他乳制品(包括酪乳或酸乳)、果汁或鸡蛋等液体代替水。它们还提供了额外的甜味、脂肪或发酵组分,以及水。牛乳面团质地细腻,风味更好,上色更为迅速,牛乳面团也可以补充完整的蛋白质。

面包改良剂:商业面包制作中通常加面包改良剂和面团调节剂,旨在减少膨胀时间、改善质地和体积。这些物质可以是强化面团的氧化剂,或削弱面筋和降低混合时间的还原剂,或可强化面团或提供其他性能如使切片更容易的乳化剂,或增加产气量的酶。

至于包装材料,有些国家使用聚乙烯膜。但许多国家则使用蜡纸或防油纸。

加工步骤

做面包有很多步骤。它们是:称量、混合、发酵、分割、搓圆、烘烤和冷却。

工厂生产面包需要以下操作程序:

(1)带气力输送系统的风筛机

面粉流送,进入风筛机,以除去杂质,然后由气力输送系统送入进料斗。

(2)搅拌

面团由面粉和酵母、糖、盐、添加剂、酵母养料等物质制成。搅拌之后,撕下一小块面团,当撕扯面团时面团变得足够薄和非常透明。这种搅拌机有低、高速两档速度。对于大规模生产,宜用高速卧式混合机。面团在室内发酵 2~4h 后,再次进行混合。

(3)面团分切

将面团称量后分切混合好的面团。

(4)搓圆

过秤和分割之后的面块,由面团分切机的输送带传送,并由搓圆机搓圆。

(5)发酵

搓圆后的面坯由斗式输送机传送,经过约 15min 的发酵,使面筋松弛,易于处理。

(6)整形

发酵过的面块由滑槽传送,由整形机成形。

(7)醒发

架子上已经膨松的面团被送往醒发室,醒发室温度 38~40℃,湿度 80%~85%。根据制作的面包类型,通常醒发 25~40min。醒发后,在面团的表面刷蛋液、水、牛乳或蛋清。

(8)烘烤

发酵好的面团在预热好的烤炉中烘烤。烘烤温度 232.2~260℃,烘烤时间 40~50min。

(9)冷却

烘烤好的产品由冷却输送带输送,在输送过程中冷却。

(10)切片和包装

冷却后的产品自动切片和包装。

加工说明

使用高蛋白的面粉效果最好,蛋白质含量太低的面粉,生产的面包体积和质地差。如果面团储藏时间过久会导致面团"短"且易碎。面粉应存放在密闭容器中,置于阴凉干燥处(湿度小于 60%)。通常,面包和蛋糕粉在 21.1℃可以保存 6 个月至一年,4.4℃可以保存 2 年。如果可能的话,全麦面粉应该冷藏或冷冻。在使用冷藏或冷冻面粉之前,让它加热至房间温度。制作面包时,精确的确定面粉的量是不可能的,面团受热、湿度、糖和面包师的高度、个性、情绪的影响。如果面粉过多,面包可能是又重又硬。如果用的太少,面包就会不膨起,结果面包体积小。

制造面团的水应具有一定温度,使面团最终的温度在 23.9~27.6℃,确切的水温取决于预定的发酵时间。因此热天应使用冷水,冷天使用温水。可将面团温度乘 2 减去面粉温度便得水温。

盐也不能忽视,因为它控制酵母的作用。不加盐的面包除了有一种非常清淡的味道,往往会过度膨胀,而且面包的质地也不同于含盐的面包。

当面包配方里含麸皮、小麦胚芽、碾碎的干小麦或小麦时,面团尽量保持湿润,因为这些成分会吸收液体,使面包偏干。可以减少揉搓的次数,避免用这些具锋利边缘的物质切割面筋。随着面包粉中全麦面粉比例的增加,发酵时间也随之增加。

面团形成所需的时间有很大不同,这取决于温度、湿度、酵母特性、面粉特性和揉捏等因素。检测面团是否充分揉捏,用手指戳面团,面团应该回弹,有时在面团的表面形成气泡,这是面团被充分揉捏的又一个迹象。

Exercise

1. Answer questions

(1) What raw material we need for bread?

(2) Which kind of yeast is convenient to use for bead? Why?

(3) How many steps have been introduced when some plants to manufacture bread?

(4) How many fermentations are there in bread making?

(5) What are the conditions of second fermenting or proofing?

2. Fill in the blanks with the proper words

(1) It is not only _____ in all countries of the world, but it is also the most _____ form of food there.

(2) Each of which has its own characteristic in _____ , _____ of making and _____ .

(3) Wheat flour, the main raw material for bread many broadly be classified into _____ protein, _____ protein flour and _____ protein.

(4) Yeast is a living plant which requires _____ , _____ and _____ so that it can live and reproduce.

3. Translation

(1) Bread is a staple food prepared from a dough of flour and water, usually by baking.

(2) Whole meal flour can be used for bread making but bread will be coarse in the texture dark in color.

(3) After baking, remove the bread from its pan, set on a rack and let cool slowly in a draft-free place. When cooled, place in a plastic bag or plastic wrap and store at room temperature. It will last from two to seven days, depending on the bread.

Lesson 8

Reading Material

Sausage Making

A highly seasoned minced meat (such as beef, veal, pork, lamb, poultry or any combination of these meats) usually stuffed in casings of prepared animal intestine, it would become a sausage. Sausage making originally developed as a means to preserve and transport meat. Primitive societies learned that dried berries and spices could be added to dried meat. By 600—500 BC there is recordation of sausages from China, Rome and Greece. The word "sausage" is derived from the Latin word "salsus", which means salted or preserved by salting.

The procedure of stuffing meat into casings remains basically the same today, but sausages and sausage products have since evolved into a wide variety of flavors, textures, and shapes resulting from variations in ingredients. To make any sausage you need marinade, the type of marinade you choose will determine the variety of sausage you will make. Sausage processing involves a wide range of physical and chemical treatment methods, normally combining a variety of methods. We can break sausage production down into four basic process: Selecting ingredients, grinding and mixing, stuffing and thermal processing. The sausage production processing major steps and equipments as shown in the fig 8.1.

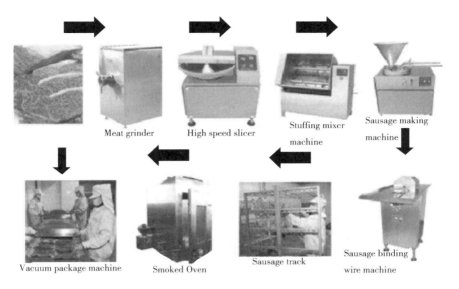

Fig. 8.1　Sausage production processing major steps and equipments

Selecting ingredients

The finished product is only as good as the ingredients it contains. For sausage, chose meat should be scraps with no skin, gristle, blood clot, or pieces of bone remaining. Meat should be fresh, high quality, have proper lean-to-fat ratio and have good binding quality. The meat should be clean and not contaminated with bacteria or other microorganisms. In other words, meat used in sausage production should be safe. Selecting spices and seasonings and combining them in proper amounts is important. They must complement each other to create a satisfying product. At the same time, there are many non-meat ingredients that are essential to the sausage making process. These non-meat ingredients stabilize the mixture, and add specific characteristics and flavors to the final product.

Curing

Curing means that meat is prepared with salt, nitrite, ascorbates, erythorbates and dozens more chemicals that are pumped into the meat. Meat is cured for preservation and flavor. Meat cured only with salt, will have a better flavor but will also develop an objectionable dark color. The curing temperature should be between $2 \sim 10℃$. Lower than $2℃$ may slow down the curing process or even halt it. Commercial producers can cure at lower temperatures because they add chemicals for that purpose. There is a temperature that cannot be crossed when curing and this is when meat freezes at about $-2℃$. Higher than normal temperatures speed up the curing process but increase the possibility of spoilage.

Factors that influence curing:

The size of the meat—the larger meat the longer curing time.

Temperature—higher temperature, faster curing.

Moisture content of the meat.

Salt concentration of dry mixture or wet curing solution—higher salt concentration, faster curing.

Amount of fat—more fat in meat, slower curing.

pH—a measure of the acid or alkaline level of the meat. (Lower pH—faster curing).

The amount of nitrate and reducing bacteria present in the meat.

The cured and fresh meat color as shown in the fig. 8.2

Fig. 8.2 Cured and fresh Meat Color

Grinding

The grinding stage reduces the meat ingredients into small, uniformly sized particle. Generally, grinding processes will vary according to the manufacturer and the nature of the product. Ideally, meat should always be chilled between 0~2℃ for a clean cut. After we are done cutting the meat, we should separate it into different groups: lean, semi-fat, and fat. The lean meat should be separated from the fat. As a rule, lean meat is ground coarsely while fatty cuts are ground very finely. By this way sausage is lean-looking and the fat is less visible. The plate selection depends greatly on the type of sausage that you decide to make.

Mixing

If the meat was previously cured, then the salt, nitrite and sugar were already added to it. Now we have to add the remaining spices. They should be mixed with cold water in a blender, and then poured over the minced meat. The water helps to evenly distribute the ingredients and it also helps soften it during stuffing. Water should not be added to uncooked sausages which will be cold smoked, slow fermented or air dried.

When mixing meat with ingredients, it is best to follow this sequence:

(1) Minced meats, starter culture, nitrite/nitrate, spices.

(2) Minced fat.

(3) Salt.

Stuffing

After the blending is complete, the blended ingredients may be bulk packaged, or they may be extruded into a casing. This process is called stuffing. And the sausage filler machine is the necessary equipment for stuffing kinds of sausages.

The casing materials may be nature or manufactured. Nature casing are the gastrointestinal tracts of cattle, sheep and hogs. Generally, hog's casing are the most suitable for home using and breakfast-type sausage. They are digestible and are very permeable to moisture and smoke. Today, however, natural casings are often replaced by collagen, cellulose, or even plastic casings, especially in the case of industrially manufactured sausages. Some forms of sausage, such as sliced sausage, are prepared without casing.

Fibrous casing are more suitable for smoked sausage and similar products because of their greater strength and the variety of size available. They are permeable to smoke and moisture and can easily be removed from the finished product. Collagen casings contain the attributes of both nature and fiber casing. They have been developed primarily for use in products such as fresh pork and pepperoni sticks. They are uniform in size, relatively strong and easy to handle. These casing are also used for the manufacture of dry sausage, because they are permeable and shrinkable.

Thermal processing

It is important that meat reaches the safe internal temperature. There are basically two methods:

①Cooking in a smoker or an oven.
②Cooking in water.

Sausage is smoked and heated in order to pasteurize it and extend its shelf life, as well as to import a smoky flavor and improve its appearance. Smoking and heating also fixes the color and cause protein to move to the surface of the sausage so it will hold its shape when the casing is removed. A few products, such as mettwurst, are smoked with a minimum of heating and are designed to be cooked at the time of consumption. Others, such as liver sausage, are cooked but not smoked.

1. Procedure for smoking sausage

After stuffing in hog casing (pre-flushed) let hang and dry. Smoke at 48.9℃ for 1 hour, 65.5℃ for one more hour, the at 76.6℃ for 2 hours or until an internal temperature of 60.6℃ is reached. Remove from smoke house and spray with hot water for 15 to 30 seconds. Follow with cold shower or dip in a slush tank until internal temperature reaches 37.8℃. Let drying for 1 to 2 hours. Place in a cooler.

2. Procedure for making cooked sausage

After stuffing the ground ingredient into an impermeable casing, put the sausage into a pan of water. Heat water to 76.7℃ and holds it there until the sausage reaches 68.3℃. A thermometer is essential for obtaining proper temperature. The water should not boil, as this will ruin the product. If you are making a sausage product using cooked meat, make sure the meat was cooked with low heat.

Vocabulary

veal 牛肉;小牛
pork 猪肉
lamb 羔羊肉;羔羊,小羊
poultry 〈集合词〉家禽
spice 香料;调味品
seasoning 调味品,佐料
ascorbate 抗坏血酸盐
erythorbate 异抗坏血酸盐(用于食物中作为抗氧化剂)

objectionable 令人不快的,令人反感的,讨厌的
grind 磨碎,嚼碎
gastrointestinal 胃与肠的
collagen 胶原质,胶原蛋白
pasteurize 用巴氏灭菌法对(牛乳等)消毒(灭菌)
mettwurst 生熏软质香肠

参考译文

香 肠 制 作

将调味好的碎肉(如牛肉、小牛肉、猪肉、禽肉或这些肉的混合物),填充到动物肠衣中它就是香肠了。香肠最初是为了保存和运输肉类而开发出来的,原始社会时期就懂得将干果和调料添加到干肉中。到公元前 600 至公元前 500 年,中国、罗马和希腊均有香肠的记载。香肠来源拉丁单词"salsus",这个单词的意思是腌制或通过腌制保存。

将肉填充到肠衣的程序基本保持与现在相同,但是香肠和香肠产品由于原料的不同已经演变成各种不同的口味,质地和形状。制作任何香肠都需要一定的腌料,所选择腌料的种类决定了所要制作香肠的种类。香肠的加工涉及许多物理和化学处理方法,通常采用多种方法相结合。我们将香肠加工分为四个基本过程:选取原料、搅碎及滚揉、灌制和热加工。

香肠制作过程的主要步骤及机器见图 8.1 所示。

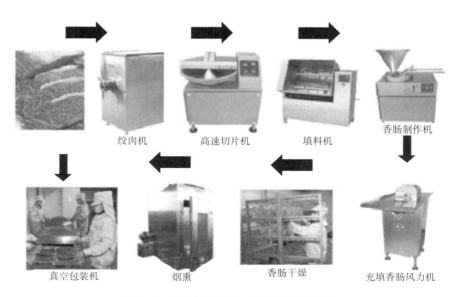

图 8.1 香肠制作过程主要的步骤及机器

原料选择

有好原料,才有好的产品。制香肠要选用没有皮、软骨、血块和残留碎骨的肉块。肉应该新鲜、质量好,有适当的瘦肉和脂肪比例以及良好的结合性能。肉类应该干净,不被细菌或其他微生物污染,也就是说,用于香肠生产的肉应该是安全的。香料和调味料的选择和适当配比是很重要的。它们必须相互补充才能制造出令人满意的产品,同时也有许多非肉类原料对香肠制作至关重要。这些非肉类成分可稳定混合物,并为最终产品增加了特殊的风味和特性。

腌制

腌制就是将盐、亚硝酸盐、抗坏血酸盐、异抗坏血酸盐和许多化学物质注入到肉中。肉腌制是为了防腐和风味。肉只用盐腌制将有较好的风味,但是也会形成令人不愉快的黑色。腌制温度应在2~10℃之间,低于2℃可能会减缓甚至停止腌制过程。工业化生产可以在较低温度下进行腌制,因为他们为此添加了一些化学试剂。在腌制过程中有一个温度是不能跨越的,就是肉在-2℃的结冰温度。高于正常温度加快了腌制过程,但也增加了变质的可能性。影响腌制的因素有:

肉的大小——肉块越大,腌制时间越长。

温度——温度越高,腌制越快。

肉的水分含量。

干混合物中盐的浓度或湿腌溶解盐的溶度越高,腌制越快。

脂肪的含量——肉中脂肪较多,腌制越慢。

pH——肉的酸碱度。(较低 pH——腌制较快)。

肉中硝酸盐和还原细菌的数量。

腌制肉和新鲜肉颜色的不同如图8.2所示。

新鲜的鸡胸肉

腌制鸡胸肉

新鲜牛肉

腌制牛肉

新鲜猪排骨

腌制猪排骨

新鲜梅花肉

腌制梅花肉

鸡肉,左边为生的,右边为腌制的

牛肉,左边为新鲜的,右边为腌制的

猪排骨,左边为新鲜的,右边为腌制的

梅花肉,左边为新鲜的,右边为腌制的

图8.2 腌制肉和新鲜肉的颜色

绞碎

绞肉阶段将原料肉绞成小而均匀的颗粒。一般来说,绞肉加工根据制造商和产品的性质而有所不同。理论上,肉应该冷却到 0~2℃,然后干净切割。在切完肉之后,把它分成不同的组:瘦肉、半脂肪和脂肪。瘦肉应与脂肪分开。一般来说,瘦肉是粗磨的,而脂肪切得很细,这样,香肠看起来瘦肉多,而脂肪就不那么明显了。筛板的选择要根据所做的香肠种类进行选择。

混合

如果肉是预先腌制的,那么已经加入了盐、亚硝酸盐和糖。现在需要添加剩下的香料。它们应该在搅拌机里用冷水混合,然后浇在肉末上。水有助于均匀分配配料,在灌制时也有助于软化。不应将水加入未经熏制、缓慢发酵或风干的生鲜香肠中。

当肉与配料混合时,最好遵循以下加入顺序:

(1)肉末,发酵剂,亚硝酸盐/硝酸盐,香料。
(2)搅碎的脂肪。
(3)盐。

灌制

混合完成后,混合原料可以散装包装,也可以将其挤压成套管。这个过程叫做填料。灌肠机是灌装各种香肠的必备设备。

肠衣材料可以是天然的和人工合成的。天然材料是牛、羊和猪的胃肠道。通常,猪肠衣最适用于家庭使用和制作早餐香肠。它们可消化,水分和烟雾易渗透进去。然而,现如今的天然肠衣经常被胶原蛋白、纤维素甚至塑料肠衣所取代,尤其是工业生产香肠。有些香肠,如切片香肠,不需要肠衣即可制备。

纤维肠衣更适用于烟熏香肠及类似产品,因为它们强度更大,而且可以形成不同的尺寸。它们更适用于烟熏和湿加工以及更容易从产品上剥落。胶原蛋白肠衣具有天然肠衣和纤维肠衣的优点。它们主要用于新鲜猪肉肠和意大利腊肠中。它们尺寸统一、相当坚韧,易于处理。这种肠衣也用于生产干香肠,因为它们具有渗透性和收缩性。

热加工

肉的内部达到安全的温度是很重要的。基本上有两种方法:

①用烟熏机或烤箱加热。
②水中加热。

香肠烟熏和加热的目的是杀菌和延长货架期,同时也可以产生烟熏风味和改善外观。烟熏和热处理可以改善颜色,也可以使蛋白质转移到香肠的表面,即使剥皮后还能保持形状。少数产品例如生熏软质腊肉香肠,用少量热烟熏,食用时再加热。其他产品如肝香肠,蒸煮但不烟熏。

(1)烟熏香肠加工工艺

灌制猪肠衣后进行悬挂和干燥。48.9℃烟熏 1h,65.5℃烟熏 1h 以上,76.6℃烟熏 2h,或者直到内部温度达到 60.6℃。从烟熏房移出,用热水喷淋 15~30s。然后用冷水淋浴或浸泡在冷水箱中直到内部温度达到 37.8℃,干燥 1~2h,然后冷藏。

(2)熟制香肠加工工艺

将滚揉后的物料灌装到不透水的肠衣后,放入水中,将水加热到 76.7℃并维持水温,直到香肠温度达到 63.8℃。为了获得准确的温度,温度计是很必要的。水不能沸腾,否则会破坏产品。如果用熟肉制作香肠,要确保熟肉是经低温加工的。

Exercise

1. Answer questions

(1) How many steps are there in sausage making?

(2) Which two categories is divided in the casing?

(3) How many methods are included in thermal processing?

(4) What does the nitrite used for?

(5) What is the effect in the smoking?

2. Translation

(1) We can break sausage production down into four basic process: selecting ingredients, grinding and mixing, stuffing and thermal processing.

(2) Although the actual massaging time is only about 1~3 hours, this action is continuously interrupted and meat is allowed to rest.

(3) Sausage is smoked and heated in order to pasteurize it and extend its shelf-life, as well as to import a smoky flavor and improve its appearance.

(4) 烟熏和热处理可以改善颜色,也可以导致蛋白质转移到香肠的表面因而在剥皮时还能保持形状。

(5) Sausage making is a traditional food preservation technique. Sausages may be preserved by curing, drying (often in association with fermentation or culturing, which can contribute to preservation), smoking, or freezing. Some cured or smoked sausages can be stored without refrigeration. Most fresh sausages must be refrigerated or frozen until they are cooked.

Lesson 9

Reading Material

Yogurt Making

Yogurt is a well-known fermented dairy product that form important part of the human diet in many parts of the world. Yogurt is a fermented milk product that contains the characteristic bacterial cultures *Lactobacillus bulgaricus* and *Streptococcus thermophilus*. Yogurt is made with a variety of ingredients including milk, sugars, stabilizers, fruits and flavor, and bacterial culture. Fermentation of lactose by these bacteria produces lactic acid, which acts on milk protein to give yogurt its texture and characteristic tart flavor. The lactic acid lowers the pH, makes it tart, causes the milk protein to thicken and acts as a preservative since pathogenic bacteria cannot grow in acid conditions. All yogurt must contain at least 8.25% solids not fat. Full fat yogurt must contain not less than 3.25% milk fat, low fat yogurt not more than 2% milk fat, and nonfat yogurt less than 0.5% milk. Depending on the system of manufacture and the nature of the coagulum, yogurt may be classified as being of two main types, set or stirred. The manufacturing of yogurt as shown in the fig. 9.1.

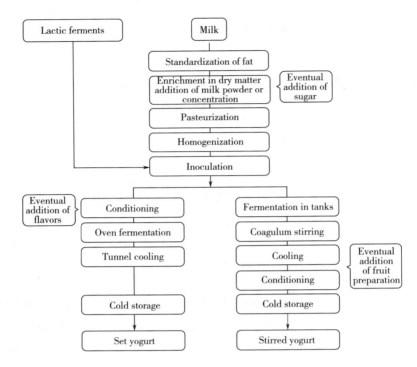

Fig. 9.1 The manufacturing of yogurt

Raw materials

The type of milk used depends on the type of yogurt – whole milk for full fat yogurt, low fat milk for low fat yogurt, and skim milk for nonfat yogurt. Other dairy ingredients are allowed in yogurt to adjust the composition such as cream to adjust the fat content, and nonfat dry milk to adjust the solid content. The solid content of yogurt is often adjusted above the 8.25% minimum to provide a better body and texture to the finished yogurt.

To modify certain properties of the yogurt, various ingredients may be added. To make yogurt sweeter, sucrose (sugar) may be added at approximately 7%. For reduced calorie yogurt, artificial sweeteners such as aspartame or saccharin are used. Cream may be added to provide a smoother texture. The consistency and shelf stability of the yogurt can be improved by the inclusion of stabilizers such as food starch, gelatin, locust-bean gum, guar gum and pectin. These materials are used because they do not have a significant impact on the final flavor.

To improve taste and provide a variety of flavors, many kinds of fruits are added to yogurt. Popular fruits include strawberries, blueberries, bananas, and peaches, almost any fruit can be added. Beyond fruits, other flavorings are also added.

Including vanilla, chocolate, coffee, and even mint. Sweeteners, flavors and fruit preparations are used in yogurt to provide variety to the consumer.

The processing of yogurts can be broken down into the following steps: adjusting milk composition and blending, pasteurization, homogenization, culturing and cooling, add flavors and fruits, packaging and storage. Each step is extremely important in the process, and strict attention to detail must be taken.

Adjusting milk composition and blending

Milk composition may be adjusted to achieve the desired fat and solids content. Often dry milk is added to increase the amount of whey protein to provide a desirable texture. Ingredients such as stabilizers are added at this time.

Homogenization

Homogenization is a process in which the fat globules in milk are broken up into smaller, more consistently dispersed particles. Homogenization is accomplished using a homogenizer or viscolizer. In this machine, the milk is forced through small openings at a high pressure and fat globules are broken up due to shearing forces. In the manufacture of yogurt and other dairy products, it is common to homogenize mixes at approximately 63℃, with a total pressure between 7 and 10MPa in the 1st stage and 3MPa in the 2nd stage, or alternatively, 7MPa (1000 psi) 1st stage and 3MPa 2nd stage. The homogenizing as shown in the fig. 9.2.

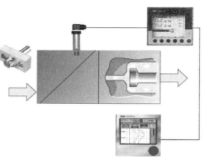

Fig. 9.2 Homogenizing

Pasteurization

As with any other dairy product, the purpose for pasteurization is to heat treat milk to eliminate pathogenic bacteria. In addition, it is very important to denature the proteins to attain the highest level of functionality from the milk proteins. Pasteurization also aids in the hydration of the stabilizers and dry ingredients that were added during blending, as well as adding a pleasant cooked flavor. The three main types of pasteurization are vat method (low temperature long time LTLT, 80℃ for 30 min), high temperature short time (HTST, 80~88℃ for 18~50s depending upon length of holding tube) and ultra-high temperature (UHT, 138℃ for 2~4s). The pasteurizing as shown in the fig. 9.3.

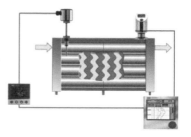

Fig. 9.3 Pasteurizing

Fermentation

When pasteurization and homogenization are complete, the milk is cooled to between 43~46℃ and the fermentation culture is added in a concentration of about 2%. It is held at this temperature for about 3~4 hours while the incubation process takes place. During this time, the bacteria metabolizes certain compounds in the milk producing the characteristic yogurt flavor. An important byproduct of this process is lactic acid. As soon as the pH value reaches 4.2 to 4.5, the yoghurt must be cooled from the incubation temperature to 15 to 22℃, to stop acidification. The fermentation as shown in the fig. 9.4.

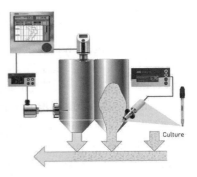

Fig. 9.4 Fermentation

Depending on the type of yogurt, the incubation process is done either in a large tank of several hundred gallons or in the final individual containers.

(1) Stirred yogurt

This type of yogurt is fermented in a large vat. After fermentation, the content is cooled. The content is stirred with fruit and flavor and then poured into the final selling containers. The products are stored at refrigeration temperatures.

(2) Set yogurt

Set yogurt, also known as French style, is allowed to ferment directly in a container for sale. After fermentation, the product is cooled and stored at refrigeration temperatures.

Cooling

After incubation, the yogurt is immediately cooled to storage temperature. The cooling as shown in the fig. 9.5.

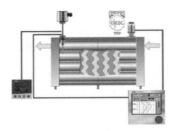

Fig. 9.5 Cooling

Yogurt and other cultured dairy foods are excellent sources of high-quality protein and certain B-complex vitamins and minerals. In addition to the nutritional importance of yogurt, it has been found to be of value in maintaining a balance in the intestinal microbial flora.

Vocabulary

lactobacillus bulgaricus　保加利亚乳杆菌
streptococcus thermophilus　嗜热链球菌
whey　乳清
lactalbumin　乳清蛋白
lactoglobulin　乳球蛋白
tart　尖刻的；酸的
pathogenic　致病的
aspartame　天冬氨酰苯丙氨酸甲酯
saccharin　糖精
gelatin　凝胶，白明胶
locust-bean gum　刺槐豆胶
guar gum　瓜尔豆胶

pectin　胶质
vanilla　香草；香草精
mint　薄荷
pasteurization　巴氏杀菌法
pathogenic bacteria　病菌，病原菌
denature　使变性；使变质
hydration　水合，水作用
homogenizer　均质器，高速搅拌器
viscolizer　均质机，匀化器
incubation　孵化；孵育
coagulum　凝结物，凝固物

参考译文

酸 乳 制 作

众所周知，酸乳是一种发酵乳制品，它构成了世界许多地方人类饮食的重要组成部分。酸乳是具有保加利亚乳杆菌和嗜热链球菌细菌培养特征的一种发酵乳制品。酸乳是由牛乳、糖、稳定剂、水果和香料，以及细菌培养物等多种原料制成。这些细菌发酵乳糖产生乳酸，作用于乳蛋白，使酸乳具有质地和风味。乳酸降低了pH，使牛乳变酸，引起牛乳蛋白质变得浓稠和作为防腐剂，因为致病细菌在酸性条件下不能存活。所有酸乳必须至少含有8.25%非脂固体，全脂酸乳必须含有不少于3.25%的乳脂肪，低脂酸乳乳脂肪不超过2%，脱脂酸乳乳脂肪少于0.5%。根据生产系统和凝胶性质，酸乳可以分为两个主要类型，凝固型或搅拌型。酸乳制作过程如图9.1所示。

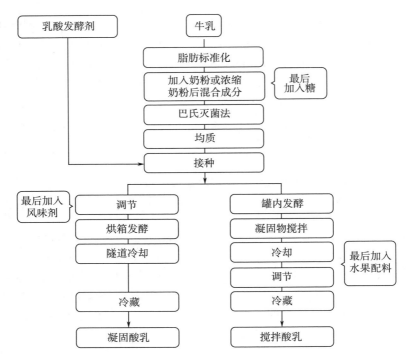

图 9.1　酸乳制作

原材料

使用的牛乳类型取决于酸乳的类型,全脂牛乳生产全脂酸乳,低脂牛乳生产低脂酸乳,脱脂牛乳生产脱脂酸乳。允许其他乳原料调整酸乳成分,如奶油调节脂肪含量,脱脂乳粉调整固体含量。酸乳中的固体含量通常被调整到 8.25% 以上,从而为成品酸乳提供更好的质地。

要改变酸乳的某些特性,可以添加各种成分。为了使酸乳更甜,蔗糖(糖)可添加约 7%。对于低热量的酸乳,使用人工甜味剂如阿斯巴甜和糖精。添加奶油,可以提供更平滑的质构。添加淀粉、明胶、刺槐豆胶、瓜尔豆胶、果胶等稳定剂,可以提高酸乳的稠度和货架期稳定性。使用这些材料,是因为它们对最终风味没有显著影响。

为了改善口味和提供多种风味,许多水果添加到酸乳中。受欢迎的水果包括草莓、蓝莓、香蕉和桃子,几乎所有的水果都可以添加。除了水果,其他的调味料也加。这些可以包括香草、巧克力、咖啡甚至薄荷等东西。酸乳中使用甜味剂、香精和水果制剂为消费者提供多样性。

酸乳的加工可以分为以下几个步骤:调整牛乳成分和混合、杀菌、均质、培养、冷却、添加香料和果肉、包装和储藏。在这个过程中,每一个步骤都是非常重要的,必须严格注意细节。

调整牛乳成分和混合

可以调整牛乳成分以达到所需脂肪和固体含量。通常添加乳粉以增加乳清蛋白的含量来提供理想的质地,此时加入稳定剂等物质。

均质

均质化是将牛乳中的脂肪小球分解成更小、更一致的分散颗粒的过程。使用均质机完成均质。在均质机中,牛乳被迫在高压力下通过小孔,由于剪切力脂肪球被破碎。在酸乳和其他乳制品的生产中,常见的均质是第一阶段约63℃,总压力为7~10MPa之间;第二阶段,3MPa。或者,第一阶段7MPa和第二阶段3MPa。均质如图9.2所示。

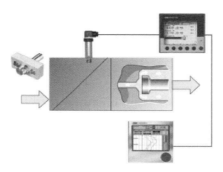

图9.2 均质

巴氏杀菌

和其他乳制品一样,巴氏杀菌的目的是加热牛乳以消除致病细菌。另外,它对牛乳蛋白质变性达到最高功能很重要。巴氏杀菌也有助于在混合过程中添加稳定剂和干物质的水化,以及添加一种令人愉快的烹饪味道。巴氏杀菌主要有三种类型:大缸方法(低温长时间巴氏杀菌法 LTLT,80℃ 30min),高温短时间巴氏杀菌法(HTST,80~88℃ 18~50s 根据保温管的长度确定时间)和超高温瞬时杀菌(UHT,138℃ 2~4s)。巴氏灭菌如图9.3所示。

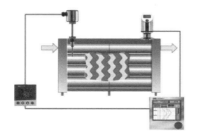

图9.3 巴氏灭菌

发酵

当巴氏杀菌和均质完成后,将牛乳冷却至 43~46℃,加入约 2% 的发酵剂。在这温度下保持约 3~4h,进行乳酸菌繁殖。在这期间,细菌代谢牛乳中某些化合物,产生酸乳的风味特征,这个过程的一个重要副产品是乳酸。当 pH 达到 4.2~4.5 时,酸乳就必须从繁殖温度冷却到 15~22℃,停止酸化。发酵如图 9.4 所示。

图 9.4　发酵

根据酸乳的种类,发酵过程是在几百加仑的大罐中或在最后的单个容器中进行的。

(1) 搅拌型酸乳

这种酸乳是在发酵罐中发酵的,发酵之后进行冷却,和水果、香精一起进行搅拌,然后装入最后的市售包装盒,产品在冷藏温度下储藏。

(2) 凝固型酸乳

凝固型酸乳,也被称为法式酸乳,直接在待售的容器中发酵。发酵后,产品在冷藏温度下储存。

冷却

发酵后,酸乳立即冷却至储存温度。如图 9.5 所示。

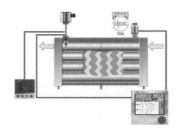

图 9.5　冷却

酸乳和其他发酵乳制品都是优质蛋白质和某些 B 族维生素及矿物质的良好来源。酸乳除了其营养价值,还发现了其在维持肠道菌群平衡的重要性。

Exercise

1. Answer questions

(1) What is yogurt?

(2) How many steps are there in the yogurt manufacturing process?

(3) Which two categories is divided in yogurt?

(4) Is yogurt one of many types of fermented milk which is consumed around the world?

(5) Which kind of bacteria is inoculated in yogurt production?

2. Translation

(1) Milk composition may be adjusted to achieve the desired fat and solids content.

(2) Pasteurization also aids in the hydration of the stabilizers and dry ingredients that were added during blending, as well as adding a pleasant cooked flavor.

(3) When pasteurization and homogenization are complete, the milk is cooled to between 43 ~ 46℃ and the fermentation culture is added in a concentration of about 2%.

(4) 均质化是将牛乳中的脂肪小球分解成更小、更一致的分散颗粒的过程。

(5) Freshly drawn bovine milk has a pH of about 6.6. At this pH, casein (a milk protein) exists as a colloidal of a calcium salt called calcium caseinate which gives milk its white colour and turbid appearance. When streptococci begin to excrete lactic acid, some of it reacts with thecalcium caseinate to form calcium lactate and free (soluble) casein. As the pH approaches 4.6, casein begins to denature (coagulate), forming a smooth semisolid curd. Such an acid curd is typical of milk fermentation by the lactic acid bacteria.

Lesson 10

Reading Material

Flour Milling

Wheat flour has been defined as the product prepared from grain of wheat by grinding or milling processes, in which the bran and germ are partly removed and the remainder is comminuted to a suitable degree of fineness.

The number of parts of flour by weight produced from per 100 parts of wheat milled is known as the four yield, or percentage extraction rate.

The central portion of the seed, the endosperm, is the main source of flour protein and starch. The germ and aleurone layer are high in oil and certain vitamins, such as vitamin E which can have some medical benefits. The seed coat, or bran, is the outer layer of the seed. An average well-matured grain of wheat is composed of approximately 82.5% endosperm, 15% bran and 2.5% germ.

Although the wheat grain contains about 82.5% of endosperm which is required for flour, it is never possible to separate it exactly from the 17.5% of bran, aleurone and germ and thereby obtain a flour of 82.5% extraction rate. The mechanical limitations of the milling process are such that in practice 75% is about the limit of white flour extraction. Further increasing of extraction darken the color of the flour through inclusion of a proportion of bran, aleurone and germ. The main objective in the production of flour from wheat is to remove the endosperm from the outer bran skins in as pure a state as possible before reducing the particles into a fine flour.

On arrival at the mill, the wheat will contain a certain percentage of impurities, such as sand, stone, chaff and straw. The flour mill being a modern process, has a cleaning plant for removing these impurities before grinding the wheat into flour. The methods of separation are various, taking advantage of the difference in size, shape, gravity and terminal velocity of the wheat and various impurities. Sieves are devised on the basis of size separation principle. Sieves are covered by woven steel, copper, or bronze wire, or perforated metal. When the stream of grain is passed over the sieve, the meshes being slightly larger than the greatest diameter of the grain, the smaller impurities and grain are allowed to pass through, the large impurities tailing over. To separate the impurities smaller then wheat, it is necessary to clothe the sieves with a mesh size smaller than the diameter of the smallest grains, the throughs in this instance being small impurities, the overtails being the wheat. Employing other specific principle, people invented many machines to separate various impurities.

Besides removing foreign impurities and badly damaged kernels, the wheat is then needed to be conditioned, or tempered by the addition of water, to the optimum moisture content for milling. As temper moisture is increased, flour color improves but flour yield decreases. Wheat hardness is important in determining temper moisture, the optimum for soft wheat being up to 2% lower than for hard wheat.

In the flour processing procedure, there are four operationally distinct systems: break system, grading system, purification system, reduction system.

Break system: Milling breaks open the seed, scrapes off as much of the bran as possible. Break system consists of four or five "breaks" stages, each followed by a sieving stage. The first break is fed with the whole grain, the subsequent breaks with the break stock from the preceding stage. The objective in the first break is to open the whole grain, and in subsequent breaks to scrape the endosperm off the bran coats. Each break rollermill is equipped with a pair of rolls, usually 254 mm in diameter and up to 1016 mm. The width of the narrow gap, "nip", between the rolls can be adjusted to allow precision grinding. The rolls rotate in opposite directions so that the surfaces of both rolls are entering the nip in the same direction. One of the rolls rotates at a faster speed than the other, the speed differential between the two generally being 2.5 : 1 for break rollermills.

Grading system: Grading is an operation classifying mixtures of semolina, middlings and dunst into fractions of restricted particle size range. The ground material leaving each break roll passes to the sieving system where sifting machines separate the mixture of particles according to size. The coarse co-product from the break system is called bran and the finer bran-like material from the purification and reduction systems is known as shorts.

Purification system: Purifying means separating mixture of bran and endosperm particles, according to their terminal velocity and particles size, by means of air current and sieves. Purification system consists of a long, narrow, sieve set at a slight angle to the horizontal, enclosed by a hood or cover. The hood, divided into sections to correspond with the sieve sections, is connected to the suction side of a fan so that air, controlled by valves, is drawn up through each sieve section. Stock to be purified is fed onto the head end of the sieve and is moved by the vibrating motion down towards the tail section. Particles of endosperm, decreasing in purity from head to tail sieve sections, fall through the respective sieves, against the direction of the air current (particles of endosperm being relatively dense and having high terminal velocity). Light particles of bran are lifted by the air current. Particles intermediate in density and terminal velocity, are not lifted, nor do they fall through the sieves, but remain floating on the sieve, and are eventually discharged as overtails.

Reduction system: The modern rollermilling process for making flour is described as "gradual reduction process" because the grain and its parts are broken down in a succession of relatively gentle grinding stages, using rollermills in each. In the first stage the grain is opened up, in the subsequent stages parts of the grain selected from the products of preceding grinds are reground. Reduction system consists of 8 to 16 grinding stages (according to the type of wheat being milled, the usual moisture content, flour quality and extraction rate required).

Vocabulary

mill   磨坊;制造厂
bran   糠,麸
germ   幼芽,胚芽,胚原基
comminute   弄碎,把……弄成粉末
aleurone   糊粉
sieve   筛,滤网
perforate   打孔
mesh   网眼,筛孔,网络
temper   调和,调节
optimum   最适宜(的)
hopper   (磨粉机等的)漏斗,送料斗,加料斗
shoe   喂料器

regulate   管理,控制;调整
stationary   静止的,不动的
scrape   刮,擦
preceding   在前的;在先的
differential   差别的,特定的,微分的
semolina   砂子粉,皮磨粗粒
middlings   皮磨细粒,细麸
dunst   粗粉
coarse   粗糙的,粗鄙的
shorts   细麸
horizontal   地平的;水平的
hood   风帽,头巾;机罩
suction   吸入,吸力,抽气,抽气机,抽水

泵,吸引	eventually 终于;最后
valve 阀,阀门;电子管	discharge 排出,释放
vibrate 颤动	overtail 筛上物

参考译文

面 粉 加 工

小麦面粉一直被定义为小麦经过研磨而得到的产品。在研磨过程中,小麦的表皮和胚部分被去除,剩下的部分被粉碎到合适的细度。

按质量计,每100份小麦经过研磨所生产出来的面粉数量叫出粉率,或者叫(面粉)提取率。

小麦的中心部分,即胚乳,是面粉蛋白和淀粉的主要来源。胚和糊粉层含有很高的油和一些维生素,如具有医学作用的维生素E。小麦种子的外衣,或者说麸皮,处于种子的外层。一般成熟的小麦由将近82.5%的胚乳、15%的麸皮和2.5%的胚所组成。

尽管小麦含约82.5%的生产面粉所必需的胚乳,但人们却无法将其从总量占17.5%的麸皮、糊粉层和胚中分离出来,从而获得82.5%的出粉率。由于机械的局限性,实际上,75%出粉率已是极限,再增加出粉率会由于面粉中混入一部分麸皮、糊粉层和胚而使粉色变暗。从小麦生产面粉的主要目的就是小麦在研磨成细粉之前人们需要尽可能纯地把胚乳从小麦外表皮中分离出来。

小麦在到达面粉厂时将包含一定量的杂质,例如:沙子、石子、草屑、麦秆等。使用现代工艺的面粉厂在小麦制粉之前,有一个清理车间以去除这些杂质。分离杂质的方法很多,这些方法利用了小麦跟各种杂质的颗粒大小、形状、密度和悬浮速度等不同特性。筛子的设计是基于颗粒大小不同的分离原理,筛面是用钢丝、铜丝或者青铜丝织成,或用金属板打孔做成。当粮食流过筛面时,由于筛网网眼比最大直径的粮食颗粒还要稍大一点,因此,小的杂质和粮食被允许通过,而大的杂质就被截留了。为了分离比麦粒小的杂质,筛面有必要被织成网眼比最小的麦粒还要小,这样,筛下物就是小的杂质,筛上物就是小麦。利用其他的具体原理,人们还发明了许多机器来分离各种各样的杂质。

除了去除外来杂质和严重受损的籽粒,小麦还需要进行调理,或通过加水进行调质,以达到碾磨所需的最佳含水量。随着调质水分的增加,面粉颜色会改善,但是面粉产量会下降。小麦硬度在确定调质水分含量时非常重要,软质小麦的最佳硬度比硬质小麦低2%。

在面粉加工过程中,有四种不同的操作系统:皮磨系统,分级筛理系统,清粉系统和心磨系统。

皮磨系统:磨辊磨破种子,尽可能多地将胚乳从麸皮上剥刮下来。皮磨系统包括四道或者五道皮磨,每一道后都紧跟着筛理。第一道皮磨喂入整粒小麦,后道皮磨喂入上一道皮磨的筛上物。第一道皮磨的目的是剥开整粒小麦,接下来各道皮磨的目的是把胚乳从麸皮层剥刮下来。每一道皮磨机都配有一对辊子,通常直径为254mm,长度达到1016mm。机距能够被调整到精确的研磨要求。两辊按相反方向旋转,这样两辊表面就都按相同的方向进入辊间。其中一个磨辊旋转速度比另一个快,皮磨系统的速比通常是2.5:1。

分级筛理系统:分级筛理就是把皮磨粗粒、皮磨细粒和粗粉按照限定的粒度范围进行归类和分离的一种操作。皮磨物料进入筛理系统,在这儿,筛理设备将根据颗粒大小分离出各种颗粒范围的物料。从皮磨来的筛上物,被称为粗麸,而从清粉和心磨来的筛上物,粒度小,被称为细麸。

清粉系统:清粉意指根据麸皮与胚乳悬浮速度的差异,用风选和筛选将它们分离出来。清粉机由一狭长的筛面组成,该筛面与水平面呈一小夹角,并沿纵向方向振动,筛面由机罩密封起来。筛面分四段,从头至尾每段筛面的筛孔越来越大。与每段筛面相对应,机罩也分成数段,并与风机的吸入口相联,使空气经筛面抽入风机,风量由阀门调节。需清粉的物料由清粉机进料端进入,靠振动向出口端移动。从头至尾的各段筛面上,胚乳的纯度越来越小,胚乳逆气流方向(胚乳颗粒密度大,因而悬浮速度较高)穿过各段筛面。细薄片状的麦皮和麸屑则被气流带走。带有大量胚乳的麦皮,其密度、悬浮速度处于中等,它们既未被空气流带走,也未穿过筛孔,而是浮在筛面上,最后作为筛上物排出。

心磨系统:现代制粉的辊式磨粉工艺被称作"逐步研磨工艺",因为麦粒及其中部分在辊式磨粉机内经相对温和的研磨而成粉。在第一阶段将麦粒剥开,在以后各道研磨中则将前道磨下物料中分离出来的制品再进行研磨。心磨系统由8~16道磨组成(根据研磨的小麦类型、通常的小麦含水量、面粉质量和出粉率要求确定道数)。

Exercise

1. Answer questions
(1) What is wheat flour?
(2) What is the flour yield?
(3) What is the structure of an average well-matured wheat?

(4) What is the principle of removing the impurities?

(5) How many operationally distinct systems are there in the flour milling?

2. Translation

(1) Employing other specific principle, people invented many machines to separate various impurities.

(2) The hood, divided into sections to correspond with the sieve sections, is connected to the suction side of a fan so that air, controlled by valves, is drawn up through each sieve section.

(3) Particles intermediate in density and terminal velocity, are not lifted, nor do they fall through the sieves, but remain floating on the sieve, and are eventually discharged as overtails.

(4) 分离杂质的方法很多,包括利用小麦跟各种杂质的颗粒大小、形状、密度和悬浮速度等不同特性的方法来分离。

(5) The successful experiences of the People's Republic of China in solving the problem of feeding the whole population can be summarized as follows: It has always stuck to the principle that agriculture is the basis of the national economy, giving top priority to agriculture in national economic development. It has made the increase of grain production the key point in rural economic work, making all possible efforts to ensure a steady increase in total grain output. It has carried out the reform of rural relations of production, including implementation of the policy of the household contract responsibility system and taking the policy of integrating centralization and decentralization.

Lesson 11

Reading Material

Rice processing

Rice is one of the leading food crops of the world, and is produced in all continents. Rice was an important food even before the dawn of written history. The peoples of different countries have varying preferences for types of rice: round-grain rice is preferred in Japan, Korea and Puerto-Rico, possibly because the cooked grains are more adherent, whereas long-grained rice (e. g. Patna rice) is preferred in the USA. People in most countries prefer white rice, but in India and Pakistan the preference is for red, purple or blue strains.

The purpose of rice milling is to remove the hulls and to produce a milled rice with minimum of breakage and with a minimum of impurities, such as weed seeds in the final product.

Combine-harvested rice generally has a moisture content of abort 20% and the grain must be dried immediately to about 12% for storage.

Rice is consumed mostly in the form of whole kernels (polished rice), and accordingly the processing of the paddy, or rough rice (the threshed grains with adherent hulls), is designed to give a high yield of unbroken kernels. The total yield of processed rice is made up of the "head yield", viz the yield of unbroken kernels, plus the yield of broken kernels. The market value of whole kernels is greater than that of broken kernels, and it is important, when drying rice, to avoid conditions that promote breakage.

For example, to maintain the quality when drying rice, both the drying temperature and the amount of moisture removed at one time are limited. During the drying process, water moves from the interior of the kernel to the periphery;

shrinkage occurs, but the outer part shrinks more than the inner. If the rate of drying is too rapid, the moisture content is reduced too much at one time, the stresses set up in the kernel due to uneven shrinkage cause cracking (checking). The checked kernels break when milled and reduce the yield of whole kernel.

In the rice processing procedure, there are three basic operations:
- Cleaning (removal of foreign matter from rough rice)
- De-hulling (removal of hulls)
- Rice milling (a process in which the bran and germ are partially or wholly removed by abrasive scouring or pearling)

Cleaning: Rough rice is first cleaned to remove foreign materials, such as straw, soil particles and weed seeds. Separation of foreign matter is based on differences in gross size, weight or density, and shape (principally length) of the impurities compared to the rough rice. Some of the machines used in rice cleaning, such as scalperator, rotex sieve, carter discs, specific gravity de – stoners, and magnets, are similar to those used for the cleaning of other cereal grains.

De-hulling: The first step in the actual milling of rice is to remove the hulls (shelling). This task will be made by a rubber roll sheller. The feeder meters a falling stream of rough rice between two closely spaced rubber rollers which are turning in opposite directions and at different speeds. As the rice passes between the rollers it is subjected to a shearing force which separates the two hulls from the brown rice. The distance between the rollers may be regulated by a hand-wheel or in the modern machines by a pneumatic mechanism which automatically separates the rolls and turns off the driving motor if there is an interruption of rice flow through the machine. The product of the sheller is a mixture of whole grain brown rice, broken brown rice, unshelled rough rice and hulls. The grain issuing from the hulling machine is sifted and aspirated to remove loosened hulls, and is then passed over a paddy or table separator which separates the kernels from the unde-hulled paddy grains.

Rice milling: The bran of the brown rice grains is removed by machines called pearlers or whiteners in which the bran is abraded by pressure between the kernels and by friction between kernels and a rough steel screen. Rice mills are of basically two types: the abrasive type and friction type. In the abrasive type, brown rice is conveyed under a slight pressure into the space between a cylindrical abrasive surfaces (a mixture of emery and silicon carbide) and screen, the bran is removed by contact against a moving rough surface. Friction type functions by rubbing one kernel against another while the rice is subjected to a slight pressure. Brown rice enters the mill through an adjustable feed gate, and a short screw conveyer propels the rice under a slight pressure into the milling chamber. A rotor with projecting ridges imparts motion to the rice within a hexagonal slotted screen to the shaft; by its rotation cause the grains to slide against each other, abrading off the bran. The product from these machines is unpolished milled rice: the outer bran layers have been removed, but not the inner layers.

The unpolished milled rice is polished in a brush machine which removes the aleurone layer and any adhering particles, and yields polished rice.

The customary methods for milling rice use abrasion to remove bran from endosperm. In SEM (also called X-M) the bran layers are first softened and then wet-milled in the presence of a rice oil solution. The separated bran has a higher protein content than that of the abraded product, is virtually fat-free, and thus it is much more stable.

In the SEM process, rice oil is applied to brown rice (de-hulled rough rice) in controlled amounts, and softening is accomplished. The bran is removed by milling machines of modified conventional design in the presence of an oil solvent-rice oil / hexane miscella. The miscella acts as a washing or rinsing medium to aid in flushing bran away from endosperm, and as a conveying medium for continuously transporting detached bran rice. The miscella lubricates the grains, prevents the temperature from vising, and reduces breakage.

The de-branned rice is screened, rinsed and drained and the solvent should be removed in two stages. Superheated hexane vapour is used to flash-evaporate the bulk of the hexane remaining in the rice, and the rice is subjected to a flow of inert gas which removes the last traces of solvent.

The bran-oil miscella slurry is pumped to vessels in which the settles, and is then separated centrifugally, while being rinsed with hexane to remove the oil. The last traces of solvent are removed from the bran by flash de-solventizing, and the bran is cooled.

Advantages of the SEM process over the conventional milling process are:

1. An increase of up to 10% in head rice yield;

2. A decrease in the fat content of the rice, which improves its storage life;

3. An increase in stability of the bran product, which has potential application in breakfast cereals, baby foods and baked goods;

4. A yield of 2kg of rice oil from each 100kg of unmilled rice.

Vocabulary

projecting 突出的,伸出的
shelling 去壳,去皮,砻谷
shaft 轴,转轴
ridge 背脊,山脊
abrasive 研磨剂,研磨的
impart 传递、传授
broken 碎米(粒)
polished 擦亮的,磨光,精练的
pearling 碾,削
thresh 打谷,打,颠簸;脱粒,翻滚
solution 溶液,溶解
breakage 破坏,破损,破损量
impurity 杂质,混杂物

straw 稻草,麦秆,茎秆
detached 分开的,分离的
trace 痕迹,踪迹,微量
settle 沉淀,澄清
baked 烘焙,烤
rice milling 碾米
de-hulling 脱壳
rough rice 毛谷、毛稻
brush machine 刷米机
brown rice 糙米
shearing force 剪力
polished rice 精米

参考译文

大米加工

大米是世界上主要的粮食作物之一,各大洲都有生产,早在有文字记载以前,大米就是一种重要的粮食。不同国家的人民喜食不同类型的稻米:日本、朝鲜、波多黎各的老百姓喜欢圆形稻米,可能因为圆形稻米煮熟后黏性较大,而美国人则喜欢长粒形稻米(即巴特那稻米)。世界大多数国家的人喜食白米,而在印度和巴基斯坦,人们则喜食红、紫、蓝色米。

碾米的目的是把收获、干燥了的稻谷去除外壳和皮层,生产出在最终产品中含最少的碎米和最少含杂(如杂草子)的大米。

联合收割机收割的稻谷一般含有水分20%,贮存前必须立即干燥至12%。

稻米大多是以整粒的形式(精碾米)消费的,因而毛稻(即脱粒带壳的稻谷粒)加工要得到尽可能高的整米率。稻谷加工的总出米率由整米粒,即未破碎大米的生产率和碎米率组成。整米的销售价要高于碎米,因而在干燥稻谷时避免出现引起谷粒破裂的条件是很重要的。

例如,为了确保干燥时的稻谷品质,干燥温度及每一次降水量都必须加以限

制。在稻谷干燥过程中,谷粒所含水分从内向谷粒表面移动,使之产生收缩,但谷粒表面收缩大于内部收缩。若干燥过快,一次降水量太多,由于收缩不均而产生的应力会使谷粒破裂(爆腰),碾米时破裂的爆腰粒会使整米率降低。

稻谷的加工过程包括三个基本操作部分:
- 清理(除去稻谷中的杂质)
- 脱壳(除去稻壳)
- 碾米(一种用擦离或碾削的方法除去部分或全部米糠和胚的工艺过程)

清理:毛谷先进行清理去除杂质,如茎秆、小泥块和杂草子。分离杂质是根据毛谷和杂质在尺寸、重量或密度以及形状(主要以长度)上的差别。稻谷清理中用的一些设备如初清筛、回转筛、卡脱碟片精选机、密度去石机、磁铁等与其他谷物的清理设备类似。

脱壳:实际碾米中的第一步是去除外壳(砻谷)。脱壳是由橡胶辊筒砻谷机完成的,喂料辊定量地把下落的谷流送入两个紧靠着的橡胶辊筒之间,这两个橡胶辊旋向相反,且转速不同,当稻谷通过两辊,谷粒会受到一个剪力作用,剪力使两片外壳和糙米分离。辊间的距离可由手轮调节,一些较先进的砻谷机则采用气动机构进行调节。如果通过机器的谷料有断料时,气动机构能自动分开辊筒并切断驱动电动机,砻谷机的产物是整粒糙米、碎糙米、未脱谷稻谷和稻壳组成的混合物。砻谷机出来的谷粒要经筛理与吸风以除去碾下的谷壳,然后送入谷糙分离筛将糙米与未脱壳的毛稻谷粒分离开来,后者送入砻谷机重新脱壳。

碾米:糙米的米糠用碾米机去除。在碾米机中,米糠由于谷粒与谷粒之间的压力及谷粒与粗糙的钢质筛面之间的摩擦力而被擦掉。碾米机基本上有两种类型:碾削型和擦离型。碾削型米机,糙米在较小压力下送入碾辊(金刚砂和碳化硅的混合物)和米筛间的碾白室,皮层由与籽粒接触的、粗糙的辊面去除;而擦离型米机是在压力作用下,籽粒与籽粒间的相互摩擦达到擦离作用的。糙米通过可调节的喂料门进入米机,螺旋推进器把米粒在轻微压力下推入碾白室,碾辊凸出的碾脊把运动传递给六边形的米筛和转轴间的米粒,碾辊的转动,势必引起籽粒相互滑动,磨去皮层。碾米机出来的稻米由于仅去除了外皮而未碾去内皮而称为粗碾米。

粗碾米在刷米机中再进行精碾,除去糠粉层及其他微粒,即生产出精碾米。

传统的碾米方法是用擦离法将米糠从胚乳上除掉。而在大米溶剂提取法(也称 X-M 法)的碾米过程中,先将米糠软化再在米糠油溶液存在下进行湿碾。这样分离的米糠比擦离碾米的米糠蛋白质含量高,且不含脂肪,因而非常稳定。

在大米溶剂提取法工艺中,在糙米中加入一定量的米糠油,完成米糠的软化。然后在米糠油、己烷混合液的存在下用经过改进的传统碾米机碾下米糠。

混合液作为冲洗剂有助于米糠与胚乳的分离,也作为运输媒介连续输送从胚乳上脱离下来的米糠。混合液还可润滑谷粒,防止升温,减少碎粒。

去糠的大米要经筛理、冲洗,而溶液去除要分两个阶段进行,先将过热己烷蒸气通入稻米,快速蒸发出大米中留存的大量己烷,再将大米通入惰性气体除去残留在大米中的微量己烷。

糠-油混合液泵入一容器内,米糠沉淀,然后离心分离,同时用己烷冲洗除去糠油。米糠中残存的最后一点溶液用快速脱溶法除去,再冷却。

大米的溶剂提取法与传统碾磨法相比有以下优点:

1. 整米率可增加10%;
2. 大米中脂肪含量下降,储存时间延长;
3. 米糠产品的稳定性增加。米糠在早餐谷物、儿童食品及烘焙食品方面有很大的应用潜力;
4. 100kg 糙米可出 2kg 油。

Exercise

1. Answer questions

(1) What are the rice processing procedure?

(2) What's the purpose of rice milling?

(3) What is the moisture content stored in rice plant?

(4) What machines are used in rice cleaning?

(5) What are advantages of the SEM process over the conventional milling process?

2. Translation

(1) For example, to maintain the quality when drying rice, both the drying temperature and the amount of moisture removed at one time are limited.

(2) The bran is removed by milling machines of modified conventional design in the presence of an oil solvent – rice oil/hexane miscella. The miscella acts as a washing or rinsing medium to aid in flushing bran away from endosperm, and as a conveying medium for continuously transporting detached bran rice.

(3) The oldest techniques of grain storage have been documented in archaeological discoveries and excavations. The ancient Greeks used large earthenware vessels that were placed in cellars or underground galleries. They also made use of specially dug holes for grain storage (pit storage). Archaeologists have discovered that the storage of food grains involved many difficulties and important production losses.

Lesson 12

Reading Material

Fruit juice Processing

A wide range of drinks can be made using extracted fruit juice or fruit pulp as the base material. Many are drunk as a pure juice without the addition of any other ingredients, but some are diluted with sugar syrup. The types of drink made from fruit can be separated into two basic types:

— Those that are drunk straight after opening.

— Those that are used little by little from bottles which are stored between use.

The former groups should not require any preservative if they are processed and packaged properly. However, the latter group must contain a certain amount of permitted preservatives to have a long shelf-life after opening. The different types of drink are classified according to the following criteria (table 12.1).

Table 12.1   **The type of fruit juice**

Type	Description
Juices	Pure fruit juice with nothing added
Nectars	Normally contain 30% fruit solids and are drunk immediately after opening. They are diluted to taste with water and may contain preservatives
Squashes	Normally contain at least 25% fruit pulp mixed with sugar syrup
Cordials	Crystal-clear squashes
Syrups	Concentrated clear juices. They normally have a high sugar content

Each of the above products is preserved by a combination of natural acidity, pasteurisation and packaging in sealed containers. Some drinks (syrups and squashes) also contain a high concentration of sugar which helps to preserve them.

For all the fruit based beverages, the first stage is the extraction of juice or pulp from the fruit. The following are the key manufacturing stages: selection and preparation of raw material, juice extraction, filtration (optional), batch preparation, pasteurisation, filling and bottling.

Any fruit can be used to make fruit juice, but the most common ones include pineapple, orange, grapefruit, mango and passion fruit. Some kinds of juice, such as guava juice, are not filtered after extraction and are sold as fruit nectars.

Preparation of raw material

Select mature, undamaged fruits. Any fruits that are mouldy or under-ripe should be sorted and removed. Wash the fruit in clean water. It may be necessary to chlorinate the water by adding 1 tablespoon of bleach to 5L of water. Peel the fruit and remove stones or seeds. If necessary, chop the fruit into pieces that will fit into the liquidiser or pulper. Remember that at this stage, you are exposing the clean flesh of the fruit to the external environment. Make sure that the utensils are clean. Do not leave the cut surfaces exposed to the air for long periods of time or they may start to turn brown and this will discolour the juice. The fruit pieces can be placed in water that contains lemon juice (250ml lemon juice per litre of water) to stop them browning.

Juice extraction

There are several methods to extract juice depending on the type of fruit you use.

Crushing/grinding/disintegration——crushing for grapes and berries; Grinding for apples, pears and disintegration for tomatoes, peaches, mangoes, apricots. This processing step will need specific equipment which differs from one type of operation to another.

Enzyme treatment——enzyme treatment of crushed fruit mass is applied to some fruits by adding 2% ~ 8% pectolytic enzymes at about 50℃ for 30 minutes. This optional step has the following advantages: extraction yield will be improved, the juice colour is better fixed and finished product taste is improved.

Heating——heating of crushed fruit mass before juice extraction is also an optional step used for some fruit in order to facilitate pressing and colour fixing; at same time, protein coagulation takes place.

Pressing——pressing to extract juice.

Juice clarifying——it can be performed by centrifugation or by enzyme treatment. Centrifugation achieves a separation of particles in suspension in the juice and can be considered as a pre-clarifying step. This operation is carried out in centrifugal separators with a speed of 6000~6500rpm.

Enzyme clarifying is based on pectic substance hydrolysis; this will decrease the juice's viscosity and facilitate their filtration. The treatment is the addition of pectolytic enzyme preparations in a quantity of 0.5~2 g/L and will last 2~6 hours at room temperature, or less than 2 hours at 50℃, a temperature that must not be exceeded.

Filtering

To make a clear juice, the extracted juice or pulp is filtered through a muslin cloth or a stainless steel filter. Filtration of clarified juice can be carried out with kieselgur and bentonite as filtration additive in press-filters (equipment).

Batch preparation

When the juice or pulp has been collected, it is necessary to prepare the batch according to the chosen recipe. This is very much a matter of choice and judgement, and must be done carefully to suit local tastes. Kinds of juice are sold either pure or sweetened. Fruit squashes would normally contain about 25% fruit material mixed with a sugar syrup to give a final sugar concentration of about 40%. Squashes are diluted with water prior to use and, as the bottle is opened, partly used and then stored, it is necessary to add a preservative (for example 800mg/kg sodium benzoate). Another popular product is fruit nectar, which is a sweet mixture of fruit pulp, sugar and water which is consumed on a 'one shot' basis. Essentially, these consist of a 30% mix of fruit pulp and sugar syrup to give a final sugar level of about 12%~14%.

Pasteurisation

All the products mentioned above need to be pasteurised at 80~95℃ for 1~10 minutes prior to hot-filling into bottles. At the simplest level, this may be carried out in a stainless steel, enamelled or aluminium saucepan over a gas flame, but this can result in localised overheating at the base of the pan, with consequent flavour changes.

Care is needed when producing pineapple juice due to a heat resistant enzyme in the juice. The enzyme damages skin after prolonged contact and workers should therefore wear gloves to protect their hands. The juice must be heated to a higher temperature for a longer time to destroy the enzyme (e.g. boiling for 20 minutes).

Another option is to pasteurise the juices once they have been bottled. The bottles are placed in a hot water bath which is heated to 80℃. The bottles are held in the hot water for the given amount of time until the contents reach the desired temperature. The length of time required in the water bath depends on the size and volume of the bottles (table 12.2). A thermometer should be placed in one of the bottles, which is used as a test bottle per batch, to monitor the temperature and to ensure that the correct temperature has been reached. This method of pasteurisation has benefits but also has problems (table 12.3).

Table 12.2 Pasteurisation times at 80℃ for different bottle sizes

Bottle size/litres	Pasteurisation time at 80℃/minutes
0.33	10
0.5	15
0.75	20

Table 12.3 The pros and cons of pasteurising within after bottling

Benefits	Problems
Juice is pasteurised within the bottle so the chance for re-contamination of the juice is reduced	Difficult to ensure the internal temperature of the bottles reaches the desired pasteurising temperature
No need for large stainless steel pans for pasteurisation	Require glass bottles for pasteurising

Filling and bottling

In all cases, the products should be hot-filled into clean, sterilised bottles. A stainless steel bucket, drilled to accept a small outlet tap, is a very effective bottle filler. The output can be doubled quite simply by fitting a second tap on the other side of the bucket.

After filling hot, the bottles are capped and laid on their sides to cool prior to labelling.

Vocabulary

dilute   稀释,冲淡
guava   番石榴
nectar   果肉饮料;花蜜
squash   果汁汽水;南瓜小果
cordials   浓缩果汁,饮品
crystal-clear   透明似水晶的,易懂的

passion fruit   西番莲果,百香果
chlorinate   使氯发生作用,用氯消毒
liquidiser   (通常指电动的)果汁机
pulper   碎浆机;搅碎机;(咖啡豆的)果肉采集器
flesh   肉;肉体;果肉

disintegration	瓦解;蜕变;崩溃;<物>裂变	filtration	过滤
pectolytic enzyme	果胶分解酶	kieselgur	硅藻土
coagulation	凝结	bentonite	膨润土;皂土
centrifugation	离心作用	sodium benzoate	苯(甲)酸钠
suspension	悬挂,悬浮液	enamel	搪瓷
pectic	果胶的,黏胶质的	saucepan	长柄而有盖子的深平底锅,炖锅
hydrolysis	水解	localised	局部的
viscosity	黏稠;黏性		

参考译文

果汁制作

以提取的果汁或果肉为基本原料制成各种饮料。很多果汁是纯果汁,不加任何其他成分;但有些果汁用糖浆稀释了。果汁可以分为两种基本类型:

—— 一种是打开后直接饮用的

—— 一种是储存在瓶里,慢慢饮用

如果处理和包装得当,前一组不需要添加任何防腐剂。然而,后一种必须含有一定量允许的防腐剂开瓶后才能有长的保质期。按以下标准分成不同类型的饮料(表 12.1)。

表 12.1 果汁类型

类型	描述
果汁	不添加其他原料的纯果汁
果肉饮料	通常含有 30%果肉,可以开瓶后立即饮用。用水稀释后饮用和可以含有添加剂
果汁汽水	通常含有至少 25%果浆和糖浆混合
果汁饮料	澄清透明的果汁汽水
糖浆	浓缩的透明果汁,它们通常含糖量很高

上述每一种产品的保存都需要经过自然调酸、灭菌和容器密封。一些饮料(糖浆和果汁汽水)也含有高浓度的糖,有助于保存它们。

对于所有水果饮料,第一阶段是从水果中提取果汁或果肉。以下是关键的生产阶段:原料的选择和制备,果汁提取,过滤(可选),批量准备,巴氏灭菌,灌装。

任何水果都可以用来制作果汁,但最常见的水果有菠萝、橙子、葡萄柚、芒果和百香果。一些果汁,如番石榴汁,在提取后不会被过滤,作为果肉饮料出售。

原材料的准备

选择成熟的水果。任何发霉或熟透的水果都应加以分类和去除。在干净的水中清洗水果。可能需要用氯消毒,将 1 汤匙漂白剂加到 5L 水中。水果去皮,去核或种子。如有必要,将水果切成小块,放入果汁机或碎浆机中。记住,在这个阶段,你将水果的干净果肉暴露在外部环境中,确保工具干净。不要把切割表面长时间暴露在空气中,否则它们可能会褐变,这将使果汁变色。水果片可以放在含有柠檬(每升水 250mL 柠檬汁)的水中,以阻止它们褐变。

果汁的提取

有几种方法可以根据你使用的水果的种类来提取果汁。

破碎/磨碎/解体——破碎葡萄和浆果;磨碎苹果和梨,崩解西红柿、桃子、芒果和杏子。这个处理步骤需要特定设备,每种操作设备互不相同。

酶处理——通过添加 2%~8% 果胶酶,在大约 50℃ 处理 30min,将酶处理应用到破碎的水果块上。这一可选步骤有以下优点:萃取率提高,果汁色泽较好,成品味道改善。

加热——在果汁提取前将压碎的水果块加热,也可作为一些水果的可选步骤,以方便压榨和色泽固定;同时,蛋白质凝固也会发生。

榨汁——压榨提取果汁。

果汁澄清——可以用离心或酶的方法来进行。离心分离颗粒悬浮在果汁中,可以被认为是一个澄清步骤。该操作在离心式分离器中进行,转速为 6000~6500r/min。

酶澄清是基于果胶水解的原理,这将降低果汁的黏度,方便过滤。处理的方法是在室温下添加量为 0.5~2g/L 果胶酶持续 2~6h,或 50℃ 下小于 2h,温度不得超过 50℃。

过滤

为了制作澄清果汁,榨出的果汁或果肉经过棉布或不锈钢过滤器过滤。澄清果汁的过滤是用硅藻土和膨润土作为过滤添加剂在压力过滤器(设备)中过滤。

批量准备

当果汁或果肉被收集后,有必要根据所选择的配方来批量准备。这在很大程度上是一个选择和判断的问题,必须谨慎地做,以适应当地的口味。果汁以纯果汁或甜味果汁形式销售。水果果汁汽水通常含有约 25% 的水果原料和糖浆,最终糖浓度达到 40% 左右。在使用之前,用水冲淡,当瓶子被打开,部分使用,然后储存,有必要添加防腐剂(例如 800mg/kg 的苯甲酸钠)。另一种受欢迎的产品是果肉饮料,它是一种甜的水果果肉、糖和水的混合物。从本质上讲,这些成分包括 30% 的果浆和糖浆的混合物,最终含糖约为 12%~14%。

巴氏灭菌

上面提到的所有产品需要进行巴氏杀菌,在热罐装前在 80~95℃下加热 1~10min。最简单的,可用不锈钢、搪瓷或铝锅在气体火焰上进行,但这可能会导致锅底局部过热,随之而来的味道会发生变化。

由于果汁中有一种耐热的酶,所以在生产菠萝汁时需要小心。这种酶在长时间接触后会损害皮肤,因此工人应该戴手套来保护他们的手。果汁必须加热到更高的温度,以便更长的时间来破坏酶(如煮沸 20min)。

另一个方法是果汁装瓶后再进行巴氏杀菌。瓶子放在热水浴中加热到 80℃。瓶子被放在热水中,在一定的时间内,直到内容物达到所需的温度。水浴中需要的时间取决于瓶子的大小和数量(表 12.2)。在一个瓶子里应放置用作测试每批瓶子温度的温度计,来监控温度并确保正确的温度。这种巴氏灭菌法有好处,但也有问题(表 12.3)。

表 12.2　　不同瓶子大小在 80℃下的加热时间

瓶子大小/L	80℃下巴氏杀菌时间/min
0.33	10
0.5	15
0.75	20

表 12.3　　装瓶后的优点和缺点

优点	缺点
果汁在瓶子里巴氏杀菌,所以减少了果汁被重新污染的机会	很难确保瓶内温度达到巴氏杀菌的温度
巴氏杀菌不需要大型不锈钢平底锅	要求玻璃瓶进行巴氏灭菌

灌装

在所有的情况下,产品都应该是热灌装到干净、消毒的瓶子里。一个不锈钢的水桶,钻个孔接一个小的出口龙头,是一个非常有效的灌装器。只需在桶的另一端安装第二个龙头,就可以使产量增加一倍。

热灌装后,瓶子盖上盖子,冷却,然后贴上标签。

Exercise

1. Answer questions

(1) Which fruit and vegetable food processing we can learn in this episode?

(2) How many kinds of fruit juice according to the finished product?

(3) How can be performed in juice clarifying?

(4) How many kinds of juice extraction according to the finished product?

(5) What is the temperature and time of pasteurisation for fruit juice?

2. Translation

(1) Do not leave the cut surfaces exposed to the air for long periods of time or they may start to turn brown and this will discolour the juice.

(2) Enzyme treatment of crushed fruit mass is applied to some fruits by adding 2%~8% pectolytic enzymes at about 50℃ for 30 minutes.

(3) 水果果汁汽水通常含有约25%的水果原料和糖浆,最终糖浓度达到40%左右。

(4) Fruit juice are products for direct consumption and are obtained by the extraction of cellular juice from fruit, this operation can be done by pressing or by diffusion. The technology of fruit juice processing will cover two finished product categories: juices without pulp and juices with pulp.

Unit Ⅲ
Food Preservation
食品保藏

Lesson 13

Reading Material

Food Preservation by Canning

Food which is kept to decay because it is attacked by yeasts, moulds and bacteria. The canning process, however, seals the product in a container so that no infection can reach it, and then it is sterilized by heat. Heat sterilization destroys all infections present in food inside the can. No chemical preservatives are necessary, and properly canned food does not deteriorate during storage. The principle was discovered in 1809 by a Frenchman called Nicolas Appert. He corked food lightly in wide-necked glass bottles and immersed them in a bath of hot water to drive out the air, then he hammered the corks down to seal the jars hermetically. Appert's discovery was rewarded by the French government because better preserved food supplies were needed for Napoleon' troops on distant campaigns. By 1814 an English manufacturer had replaced Appert's glass jars with metal containers and was supplying tinned vegetable soup and meat to the British navy. The next scientific improvement, in 1860, was the result of Louis Pasteur's work on sterilization through the application of scientifically controlled heat. Today vegetables, fruit, fish, meat and beer are canned in enormous quantities.

The first stage in the canning process is the preparation of the raw food. Diseased and waste portions are thrown away, meat and fish are cleaned and trimmed, fruit and vegetables are washed and graded for size. These jobs are principally done by machines. The next stage, for vegetables only, is blanching. This is immersion in very hot or boiling water for a short time to remove air and soften the vegetable. This makes it easier to pack into cans for sterilization. Some packing machines fill up to 400 cans a minute. Fruit, fish and meat are packed raw and cold into cans, and then all the air is removed. When the cans are sealed, the pressure inside each can is only about half the pressure of the outside air. This is "vacuum" packing.

During the sterilization process which follows, the cans subjected to steam or boiling water, with the temperature and duration varying according to the type of food. Cans of fruit, for example, take only 5~10 minutes in boiling water, while meat and fish are cooked at higher temperatures for longer periods. After sterilization, the cans are cooled quickly to 32℃ to prevent the contents from becoming too soft.

The process of canning is sometimes called sterilization because the heat treatment of the food eliminates all microorganisms that can spoil the food and are harmful to humans, including directly pathogenic bacteria and those that produce lethal toxins. Most commercial canning operations are based on the principle that destruction of bacteria increases 10-fold for each 10℃ increase in temperature. Food exposed to high temperatures for only minutes or seconds retains more of its natural flavour. In the flash process, a continuous system, the food is flash-sterilized in a pressurized chamber to prevent the superheated food from boiling while it is placed in a container and further sterilizing is not required. Pasteurization combined with microfiltration can be used to extend the shelf life of milk. Milk packed in sterile containers and exposed briefly to temperatures higher than those required for pasteurization may be stored unopened for months without refrigeration.

The final stages before despatch to the wholesale or retail grocer are labelling and packing the tins into boxes. Nowadays, however, labeling is often printed on in advance by the can-maker and no paper labels are then required.

Vocabulary

microorganism 微生物
decay 腐烂
yeast 酵母
mould 霉菌

bacteria 细菌
canning 罐头制作、灌装，罐头加工的
sterilization 消毒，杀菌
microfiltration 微量过滤

参考译文

食品罐藏

食物在保存过程中受到酵母、霉菌和细菌的侵染，就会腐烂变质。然而，罐装食品加工过程中，将食品密封在容器中，使得微生物无法接触到食品，随后通过加热对产品进行灭菌。加热灭菌可破坏罐头内存在的所有微生物。在没有任何必要的化学防腐剂的情况下，正确制作的罐头食品在储存期间不会变质。1809年，法国人尼古拉·阿佩尔发现了这一原理。他把食物轻轻塞进宽颈玻璃瓶里，并将瓶子浸在热水中将里面的空气赶走，然后将瓶塞压下将罐子密封起来。阿佩尔的发现得到了法国政府的嘉奖，因为拿破仑的军队在遥远的战役中需要更好的食物供应。到1814年，英国一家制造商已经将阿佩尔的玻璃罐换成了金属容器，并向英国海军供应罐头蔬菜汤和肉。下一次的科学进步是在1860年，路易·巴斯德通过科学地控制加热来进行灭菌。如今，有大量的蔬菜、水果、鱼、肉和啤酒是进行罐装的。

罐头加工的第一步是准备食品原材料。将腐败变质和不可食用的部分剔除，肉和鱼要进行清洗和修剪，水果和蔬菜需要清洗并依照大小分级，这些工作主要由机器完成。下一步只针对蔬菜进行焯水。将蔬菜在非常热或沸腾的水中浸泡一小段时间来去除空气及软化，这个过程使蔬菜更易被装进罐头进行杀菌，一些包装机器每分钟能装400罐。水果、鱼和肉的原料经冷却后装入罐头，然后除去罐内所有的空气。当罐子被密封后，内部的压力只有外部空气压力的一半，这也算一种"真空"包装。

在随后的灭菌过程中，罐头将经受蒸汽或沸水作用，灭菌温度和持续时间依据食物的种类有所不同。例如，罐装水果只需在沸水中煮 5~10min，而肉和鱼则需要在较高的温度下煮更长的时间。灭菌后，罐头被迅速冷却至32℃，以防止

内容物变得太软。

有时,罐装加工过程也称为灭菌,因为食物热处理后,所有可能导致食物腐烂变质的微生物和那些对人类有害的微生物,包括直接致病菌和产生致命毒素的微生物将全部被杀死。大多数商业罐头生产操作都基于这一原则:灭菌温度每升高10℃,对微生物增殖的破坏力就增加10倍。食物被暴露在高温下,若只有几分钟或几秒钟可更好地保持其天然风味。在连续处理系统中,食物被放在一个加压的釜内,防止过热食品在放入容器时沸腾,并且不需要进一步灭菌。巴氏杀菌结合微滤可以延长牛乳的货架期。将牛乳包装在无菌容器中,并短时暴露在比巴氏杀菌更高的温度下,可使其保存数月而无须冷藏。

在向批发商或零售商发货之前的最后阶段是贴标和装箱。然而,现在标签通常已由罐头制造商预先印在罐子上,而不再需要纸质标签。

Exercise

1. Answer questions

(1) What is the principle of canning?

(2) Why did Appert be rewarded by the French government?

(3) Why does the process of canning be sometimes called sterilization?

2. Translation

(1) Sterilization destroys yeasts, molds, vegetative bacteria, and spore formers, and allows the food processor to store and distribute the products at ambient temperatures, with extended shelf life.

(2) 如今,罐头制品已经十分普遍,尤其是啤酒和肉制品。

Lesson 14

Reading Material

Food Preservation by Drying and Dehydration

The terms drying and dehydration are applied to the removal of water from food. To the food technologist, drying refers to natural desiccation, such as by spreading fruit on racks in the sun, and dehydration designates drying by artificial means, such as with a blast of hot air. In freeze drying a high vacuum is maintained in a special cabinet containing frozen food until most of the moisture has sublimed. Removal of water offers excellent protection against the most common causes of food spoilage. Microorganisms can not grow in a water-free environment, enzyme activity is absent and most chemical reactions are greatly retarded. This last characteristic makes dehydration preferable to canning if the product is to be stored at a high temperature. In order to achieve such protection, practically all the water must be removed. The food then must be packaged in a moisture-proof container to prevent it from absorbing water from the air. For this reason a hermetically sealed can is frequently used to store dry foods. Such a can offers the further advantage of being impervious to external destructive agents such as oxygen, light, insects and rodents.

Vegetables, fruit, meat, fish and some other foods, the average moisture content of which maybe as high as 80%, maybe dried to one-fifth of their original weight and about one – half of their original volume. The disadvantages of this method of preservation include the time and labour involved in rehydrating the food before eating. Furthermore, reconstituting the dried product may be difficult because it absorbs only about two-thirds of its original water content and this process tends to make the texture tough and chewy.

Drying was used to preserve many foods in ancient times. Large quantities of fruits such as figs have been sun-dried from ancient times to the present day. In the case of meat and fish, other preservation methods, such as smoking or salting, which yielded a palatable product, were generally preferred. Dehydration is confined largely to the production of a few dried foods, such as powdered milk, soup, potatoes, eggs, yeast and powdered coffee, which are particularly suited to the dehydration method.

Present-day dehydration techniques include the application of a stream of warm air to vegetables. Protein foods such as meat are of good quality only if freeze-dried. Liquid foods is dehydrated usually by spraying it as droplets into a chamber of hot air, or occasionally by pouring it over a drum that is heated internally by steam. Dehydration of food can be accompanied by chemical treatment. It has been reported that onions kept good quality characteristics within 6 months after pretreatment with 2.5 g/kg potassium metabisulphite during dehydration.

Vocabulary

dry 干燥；干的
dehydrate 脱水
rehydrate 再水化，水合

chamber 室,膛
agent 药剂
characteristic 特征、特性、特点

参考译文

食品的干燥及脱水保藏

干燥和脱水是指将食品中的水分除去。对于食品技术人员来说,干燥指的是自然晾干,比如把水果摊在货架上放在阳光下;而脱水是用指定的人工的方法干燥,比如用热风吹干。在冷冻干燥过程中,冷冻食品被装在特殊橱柜中保持真空状态,直到食品中大部分水分升华。去除水分能够最有效地防止常见的食品腐败,因为微生物不能在无水环境中生长,酶不具有活性,大多数化学反应均严重滞后。如果产品被储存在较高的温度下,最后的这个特点使得脱水比罐装更可取。为了达到较好的防腐效果,几乎所有的水都必须被除去,然后食物必须包装在防潮的容器中以防止吸收空气中的水分。因此,密封罐常被用来储存干燥的食品。罐装还可以进一步改善储存条件,使得食品免于遭受外部的破坏,如氧气、光、昆虫和啮齿动物。

蔬菜、水果、肉类、鱼类和其他一些食物，平均含水量可能高达80%，被干燥后，其质量只有原来的1/5，体积大约也只有原来的一半。这种保存方法的缺点是在进食前需花费时间和劳动进行复水。同时，恢复干燥食品的构成可能很困难，因为它只能吸收食品原本含水量2/3的水分，而这个过程往往会使其质地变得坚硬和耐嚼。

在古代，干燥被用于许多种食物的保存。大量的水果，如无花果，从古至今都是在太阳下晒干的。而对于肉和鱼，其他的保存方法，如烟熏或盐腌可获得良好的风味，通常是首选的。脱水主要局限于一些特别适合脱水的方法进行干制的食品，如乳粉、汤、土豆、鸡蛋、酵母和粉末状的咖啡。

目前，脱水技术包括通过热风对蔬菜进行干燥。蛋白质食物，如肉类，只有在冷冻干燥时才具有良好的品质。液体食物通常是通过将其在热空气容器中喷雾成液滴进行脱水，或者将其分布在一个内部由蒸汽加热的滚筒上进行脱水。食品在脱水加工时可同时采用化学处理。有报道称在洋葱的脱水加工过程中，使用2.5g/kg的焦亚硫酸氢钾预处理后，洋葱可在6个月内保持良好的品质特性。

Exercise

1. Answer questions

(1) What is the difference between drying and dehydrating?

(2) Why does dehydration has good effect on protection against the most common causes of food spoilage?

(3) What are the disadvantages of dehydration?

2. Translation

(1) To make the loss as small as possible, foods containing vitamin C should be cooked for as short a time as possible and should be eaten as soon as they are cooked.

(2) 由于肉和鱼脱水后口感较差，所以人们更愿意用烟熏和腌渍的方法进行保存。

Lesson 15

Reading Material

Low Temperature Storage

In order to extend shelf life and maintain quality of fresh fruits and vegetables, temperature management is an important factor. Markarian et al. (2006) developed a mathematical model based on heat transfer, water vapour, temperature and other parameters for horticultural storage facilities and the results obtained correlated with those obtained for a potato storage facility.

Exposure of microorganisms to low temperatures reduces their rates of growth and reproduction. This principle is widely used in refrigeration and freezing. The microbes are not killed. In refrigerators held at 5℃, foods remain unspoilt. In a freezer at −5℃, the crystals formed tear and shred microorganisms. This may kill many microorganisms while some are able to survive, like *Salmonella spp.* and *streptococci*. For these types of microorganisms, rapid thawing and cooking are necessary. Deep freezing forms small crystals and which reduces the biochemical activity of microbes.

Blanching of fruits and vegetables by scalding with hot water or steam prior to deep freezing inactivates plant enzymes that may produce a change in colour, etc. Brief scalding prior to freezing also reduces the number of microorganisms on the food surface by up to 99% and enhances the colour of green vegetables.

Refrigeration is the cooling of space or material below the general environmental temperature. It is applied to food material for the purpose of preservation. Refrigeration is used to extend the useful life of fresh and processed food that is required to be stored or transported from one place to another. Before the advent of modern refrigeration systems perishable foods were kept in a cool environment such as cellars or buckets immersed in water. Sometimes ice from ice-making machines was used in cities to preserve foods. The advent of mechanical refrigeration systems significantly simplified the application of refrigeration to food preservation. The first patent for mechanical refrigeration was issued in 1834 in Great Britain to the American inventor, named Jacob Perkins. Although this requires access to a regular electricity supply it was one of the easiest methods for preserving food.

Freezing was used commercially for the first time in 1842, but large-scale food preservation by freezing began in the late 19th century with the advent of mechanical refrigeration. Freezing preserves food by preventing microorganisms from multiplying. The process does not kill all types of bacteria, however; those that survive reanimate in thawing food and often grow more rapidly than before freezing. Enzymes in the frozen state remain active, although they work at a reduced rate. Vegetables are blanched or heated in preparation for freezing to ensure enzyme inactivity and thus to avoid degradation of flavor. Blanching has also been proposed for fish to kill cold-adapted bacteria on the outer surface of the fish. Various methods are used to freeze meats depending on the type of meat and the cut. Pork is frozen soon after butchering but beef is hung in a cooler for several days to tenderize the meat before freezing.

Frozen foods have the advantage of resembling the fresh product more closely than the same food preserved by other techniques. Frozen foods also undergo some changes as freezing causes water in food to expand and tends to disrupt the cell structure by forming crystals. In quick freezing, the ice crystals are smaller, producing less cell damage than if a product is frozen slowly. The quality of the product, however, may depend more on the rapidity with which the food is prepared and stored in the freezer than on the rate at which it is frozen. Some solid foods that are frozen slowly, such as fish, may, upon thawing, show a loss of liquid called drip, while liquid foods that are frozen slowly, such as egg yolk, may become coagulated. Consumer-size packages of frozen food generally may weigh up to 0.9 kg. In one type of freezer used for tin packaged foods, the packages are transported mechanically on a conveyor belt through an air blast, which produces temperatures as low as $-40℃$. Another type of freezing technique, used in the freezing of concentrated orange juice, contains a secondary refrigerant, such as calcium chloride brine, as a spray-on bath for cans at temperatures of $-29℃$. In a widely used freezer called the plate freezer, the packages are put in contact with hollow metal plates containing a refrigerant and are subjected to pressure in order to increase the rate of freezing. This method of preservation is most widely used for a great variety of foods, including bakery goods, soups and precooked complete meals.

Vocabulary

shelf life　货架期
refrigeration　制冷、冷藏
perishable　易腐败的
multiply　乘、繁殖；多样的
tenderize　使嫩化、使软化
coagulate　凝结

参考译文

低温储存

为了延长新鲜水果和蔬菜的货架期，并使其保持良好品质，温度管理是一个重要的影响因素。2006年，Markarian等建立了一个数学模型，该模型基于热量传递、水蒸气、温度以及其他与园艺储藏设施相关的参数。结果显示，该模型得出的结论与他们用马铃薯存储设施试验得到的结果一致。

低温可以降低微生物的生长和繁殖速度，这一原理被应用于冷藏和冷冻。此时，微生物并没有被杀死。冰箱的温度保持于5℃时，冰箱内的食物不会变质。冰箱的温度在-5℃时，食品内水分形成的结晶会撕裂和撕碎微生物。这一过程可能会杀死许多微生物，但也有一些微生物依然能够存活，比如沙门氏菌和链球菌。对于这些类型的微生物来说，快速解冻和烹饪是必要的。深度冻结会形成小晶体，这会降低微生物的活性。

为使植物酶失活，水果和蔬菜在深度冻结之前要进行热烫处理，这一处理会使水果和蔬菜发生颜色等方面的改变。冷冻前短时热烫处理可使得食品表面的微生物数量大大减少，这一减少量高达99%，同时热烫处理可使得蔬菜的绿色加深。

冷藏是指将空间或材料的温度冷却至一般的环境温度以下。它适用于食品材料保存。冷藏被用于延长新鲜和加工食品的使用寿命，而这些食品往往需要从一个地方运输到另一个地方。在现代制冷系统出现之前，易腐烂的食物被保存在凉爽的环境中，如地窖或者浸泡在装有水的水桶中。在城市有时用制冰机制造的冰来保存食物。机械制冷系统的出现使得用于食品保鲜的冷藏应用过程大大简化。1834年，美国籍发明家雅各布·帕金斯在英国发布了第一个机械制冷机器的发明专利。尽管这需要常规的电力供应，但这是保存食物最简单的方法之一。

冷冻是在1842年被第一次商业化应用,但大规模的应用是随着19世纪晚期机械制冷出现开始的。冷冻可以使得食品免于遭受微生物的生长而导致的侵害,然而这一过程并没有杀死所有的细菌。在食品解冻后,存活下来的细菌开始大量繁殖,繁殖速度比食品冷冻前更快。冷冻状态下的酶仍然具有活性,尽管催化效率较低,因此在冷冻蔬菜前需要热烫处理,使酶失活,从而避免其风味的损失。此外,热烫处理也被应用于鱼的冷冻产品的制备,以杀死鱼表面的嗜冷性微生物。肉的种类和切割方式不同,其冷冻方法也不同。猪肉在屠宰后不久就可冷冻,而牛肉需要在冷冻前在低温环境中悬挂几日已达到嫩化肉质的效果。

相较于其他技术,冷冻保存的食品更接近于新鲜的产品。冷冻食品也会发生一些变化,因为冷冻会导致食物中的水分的体积发生膨胀,同时晶体形成会破坏食物的细胞结构。相较缓慢冷冻,迅速冷冻形成的结晶更小,因此产生的细胞损伤更小。然而,产品的质量可能更多地取决于食物的制备和冷冻前各种操作的速度,而不是冷冻本身的速度。缓慢冷冻的固体食物,如鱼,在解冻后可能会发生"滴液"现象,使得产品液体流失;缓慢冷冻的液体食物,如蛋黄,可能会凝固。消费者用的冷冻食品包装通常质量可达0.9kg。有一种用于包装食品的冷冻设备,包装好的产品通过机械传送带运输,传送带置于鼓风冷空气中,温度低至-40℃。另外一个用于浓缩型橙汁的冷冻技术,包括一种二次冷冻剂,如氯化钙盐水,被用于在-29℃下包装罐的喷淋。在被广泛使用的平板冰柜中,包装的产品与装有制冷剂的中空金属板接触,并被压实以提高冷冻速率。这种保存方法被广泛应用于各种食品,包括烘焙制品、汤和预制的全熟食品。

Exercise

1. Answer questions
(1) What will happen when a refrigerator is held at 5℃?
(2) What are the benefits and drawbacks of the crystals formed in freezers?
(3) What are the effects of blanching of fruits and vegetables?
2. Translation
(1) The rate of freezing of plant tissue is important because it determines the size of the ice crystals, cell dehydration, and damage to the cell walls.
(2) -18℃条件下,微生物的生长完全停止,同时在冷冻的过程中,酶促和非酶促的变化以极低的速度继续进行着。

Lesson 16

Reading Material

Food Irradiation

Food irradiation can help to reduce high rates of food losses, especially with respect to cereals, root crops and dried foods. The irradiation process involves exposing the food, either prepackaged or in bulk, to a predetermined level of ionization radiation. There are two classes of ionizing radiation: electromagnetic and particulate. These are γ-rays from radionuclides ^{60}Co or ^{137}Cs. The radiation dose (level of treatment) is defined as the quantity of energy absorbed during exposure. Traditionally, the dose of ionizing radiation absorbed by irradiated material has been measured in terms of rad, but recently it has been superseded by gray (Gy), which is equal to 100 rad. One Gy represents 1J of energy absorbed per kilogram of irradiated product and the energy absorbed depends on mass, density, and thickness of food. Food irradiation doses are generally characterized as low (less than 1kGy), medium (1~10kGy), and high (greater than 10kGy). Different levels of dose are required to achieve desired results for the products. The energy level used for food irradiation, to achieve any technological purpose, is normally extremely low, e. g., 0.1 or 1kGy, which would be equivalent to a heat energy of 0.024℃ or 0.24℃. The Codex Alimentarius Commission (CAD) recommended 10Gy as the maximum energy level or dose of ionizing radiation. Thus, it is a cold method of food preservation.

The potential applications of irradiation are disinfestation, shelf-life extension, decontamination, and product quality improvement.

1. Disinfestation is one of the important postharvest treatments in food processing, and chemicals are usually used for this purpose. Disinfestation, the

control of insects, in fruits can be achieved by doses up to 3kGy. A low dose of 0.15 ~ 0.50kGy can damage insects at various stages of development that might be present on the food. Irradiation can damage insect's sexual viability or its capability of becoming an adult.

2. One form of shelf-life extension is to inhibit sprouting of potatoes, yams, onions, and garlic at 0.02 ~ 0.15kGy. Another form is to delay the ripening and senescence of some tropical fruits such as bananas, litchis, avocados, papayas, and mangoes at 0.12 ~ 0.75kGy. The irradiation also extends the shelf-life of perishable products such as beef, poultry, and seafood by decontamination of spoilage microorganisms. Usually, fruits progressively lose their resistance to phytopathogens with ripening. When a low dose is used to delay ripening, a higher level of resistance is retained in fruits, and microbial development is also delayed as an added benefit.

3. Irradiation can reduce microbial load and destruction of pathogens. One form of decontamination could be the use of a low dose (1 ~ 2kGy) to pasteurize seafoods, poultry, and beef. Another form could be a higher dose (3 ~ 20kGy), such as sterilization of poultry, spices, and seasonings.

4. A higher juice yield could be obtained if fruits are first irradiated at a dose level of several kGy, thus improving product recovery. Another study showed that the gas-producing factors in soybeans could be markedly decreased with a sequence of soaking, germination, irradiation and subsequent drying of the beans. This required a dose of 7.5kGy for maximum effect. It also facilitates reduction of the need for chemicals used in food, such as nitrite and salts. Moreover, irradiation does not leave any chemical residues in foods.

Vocabulary

irradiation 照射,辐照
dose 剂量
residue 残留物,剩余物
phytopathogen 植物病原体

ripening 成熟
senescence 衰老
disinfestation 灭虫
decontamination 消除污染,净化

参考译文

食物辐射

食物辐射处理可有效减少食品的损失,特别是谷物、块根作物和干制食品。辐射过程是将预包装或散装的食品暴露在一个预先确定好的电离辐射水平下进行处理。电离辐射包括电磁和微粒两种。这些电离辐射是来自于放射性核素钴60或铯137产生的γ射线。辐射剂量(处理水平)是指在照射过程中被吸收的能量。传统上,被辐射材料所吸收的电离辐射的剂量是采用拉德(rad)进行计算的,但近年来已被戈瑞(Gy)取代,1Gy等于100 rad。1Gy表示1kg辐射产品吸收了1J的能量,而能量吸收状况取决于食品的质量、密度和厚度。食品的辐射剂量一般分为低(<1kGy)、中(1~10kGy)、高(>10kGy)三种。需要不同辐射剂量来达到预期的产品效果。为达到任何技术要求的用于食品的辐射处理的能量水平通常都非常低,比如0.1或1kGy,相当于0.024℃或0.24℃的热能。食品法典委员会(CAD)建议将10Gy作为最大能量水平或电离辐射剂量。因此,这是一种非加热的"冷"食品保藏方法。

辐射的潜在应用领域有灭虫、延长保质期、杀菌及改善产品品质。

1. 灭虫是食品加工过程中非常重要的采后处理之一,化学品常被用于此。对于水果,用3kGy剂量的辐射可实现虫害的控制。0.15~0.50kGy低剂量可以破坏出现在食品中的处于任何生理阶段的昆虫。辐射会损害昆虫的性能力或者成长过程。

2. 辐射处理延长保质期的一种形式是抑制土豆、山药、洋葱和大蒜的发芽,其处理剂量为0.02~0.15kGy;另一种形式是延缓一些热带水果的成熟和衰老,比如香蕉、荔枝、牛油果、木瓜和芒果等,它们处理剂量是0.12~0.75kGy。辐射还可通过消灭导致食品腐败的微生物延长易腐烂产品的货架期,如牛肉、家禽和海鲜等。通常,果实随着成熟将逐渐失去对植物微生物的抗性。用低剂量的辐射处理果实时,其成熟会被延迟,因此可保持较高的抗性水平,同时作为额外的益处,微生物的生长也被推迟。

3. 辐射可减少微生物的数量和病原体的致病能力。一种方法是用低剂量(1~2kGy)的辐射对海鲜、家禽和牛肉进行巴氏杀菌;另一种方法是用高剂量(3~20kGy)的辐射对家禽、香料和调味料进行杀菌。

4. 如果预先用几个 kGy 剂量的辐射对水果进行处理,则可获得更高的果汁产量,从而改善产品的回收率。另一项研究表明,通过一系列的浸泡、萌发、辐射及随后的干燥,大豆中的产气因子可明显地减少,这需要 7.5kGy 的剂量以达到最大的减少效果。辐射还有助于减少食品中亚硝酸盐和盐等化学物质的添加。此外,辐射不会在食品中留下任何化学残留物。

Exercise

1. Answer questions

(1) How does the dose of food irradiation be quantified?

(2) What are the potential applications of irradiation?

(3) What is the principle of irradiation used to extend the shelf-life of fruits?

2. Translation

(1) A great deal of the postharvest losses due to insect infestation can be controlled and minimized by irradiating foods such as grains, pulses, tubers and fruits.

(2) 辐射处理可减少水果采后损失的原理在于,该处理一方面可延迟果实的成熟,另一方面可抑制病原微生物的生长。

Unit IV
Food Additives
食品添加剂

Lesson 17

Reading Material

Overview of Food Additives

In recent years, the use of food additives has become an essential thing in food production. Food producers add some chemical products into food during the process of food production in order to improve the food quality in tincture, odour and flavour. In addition, they use food additives to store food for a long period of time. However, some cancers can be caused by food additives. For example, many meats have some additives added which will interact with gastric acid and protein. Therefore, a large amount of nitrites which can lead to chronic physiological disorder will be produced. This chapter will first expound the definition of food additives, and then discuss the importance and hazards of them. It will finally go on to introduce some familiar categories of food additives to show why they are important.

1. What are food additives

"Food additives" means any synthetic compound or natural substance put into food to improve its quality, colour, fragrance or taste, or for the sake of preservation or processing. Some food additives have been used for centuries. With the advent of processed foods in the second half of the 20th century, many more food additives have been introduced, of both natural and artificial origin.

2. The use of food additives

The food additives being used should present no risk to the health of the consumer at the levels of use. It includes the following uses:

(1) To present the nutritional quality of the food.

(2) To enhance the keeping quality or stability of food or to improve its organoleptic properties.

(3) To provide necessary constituents for foods manufactured for groups of consumers having special dietary needs.

3. Types of food additives

Food additives by functional classification (a total of 21 categories):

√ Acidity regulator

√ Anti knot agent

√ Anti foam agent

√ Antioxidant

√ Bleaching agents

√ Swelling agent

√ Chewing gum bases

√ Colorant

√ Color fixative

√ Emulsifier

√ Enzyme preparation

√ Flavor enhancer

√ Flour treatment agent

√ Membrane agent

√ Water retention agent

√ Nutritional supplements

√ Preservative

√ Stabilizer and Coagulator

√ Sweet agent

√ Thickening agent

√ Other

For example:

Flavoring: There are approximately 1100 to 1400 natural and synthetic flavoring available to food processors.

Stabilizers: These are used to keep food structure stable or unchanged.

Colorings: 90% are artificial and do not contain any nutritional value.

Sweeteners: These designed to make the food more palatable.

Preservatives: Helps maintain freshness and prevents spoilage that is caused by fungi, yeast, molds and bacteria.

Acids/Bases: A substance used to maintain or alter the pH of food. Provides a tart flavor for many fruits, widely used in carbonated soft drinks.

Some common food additives: Curcumin, Tartrazine, Allura Red AC, Potassium sorbate Sodium erythorbate, Guar gum, Aspartame, Xylitol, Sodium Saccharin.

4. Importance of food additives

Food additives carry out a variety of useful functions which we often take for granted. Foods are subjected to many environmental conditions, such as temperature changes, oxidation and exposure to microbes, which can change their original composition. Food additives play a key role in maintaining the food qualities and characteristics that consumer demand, keeping food safe, wholesome and appealing from farm to fork.

5. The hazards of food additives

The scope and quantity of use of each food additives shall be in line with the national standards. According to the national standard, the use of a certain amount of food additives is safe, but the abuse of food additives will harm the body, and even cause cancer, malformation.

There are several reasons why food additives can cause harm:

(1) Use without the approval of the state or the use of a variety of additives.

(2) The use of additives beyond the prescribed amount.

(3) The use of additives beyond the specified range.

(4) The use of industrial grade instead of food grade additives.

Food additives is a double-edged sword. It has merits and demerits. On one hand, it not only brings benefits for us, but also promotes the development of modern food industry. On the other hand, it may be used by lawless people and consequently harms our health.

In order to avoid the demerits of food additives, the state should constantly improve related laws and strengthen supervision of food production. In addition, people should raise their legal and responsibility awareness, do not pay attention only to benefit. What's more, as consumers, we should know more knowledge about food additives, food security, and know how to select proper and healthy foods. In a word, we should try our best to avoid the demerits and make full use of the merits of food additives to improve our interests.

Vocabulary

antioxidant 抗氧化剂
acidity regulator 酸度调节剂
anti knot agent 抗结剂
anti foam agent 消泡剂
bleaching agent 漂白剂
chewing gum base 胶姆糖基础剂
colorant 着色剂
emulsifier 乳化剂
food additives 食品添加剂
fungi 真菌

gastric acid 胃酸
malformation 畸形
nitrite 亚硝酸盐
organoleptic 感官的
palatable 美味的
swelling agent 蓬松剂
stabilizer 稳定剂
sweetener 甜味剂
thickening agent 增稠剂

参考译文

食品添加剂概述

近年来，食品添加剂的应用已成为食品生产中的重要环节，食品生产者在食品生产过程中会将一些化学产品添加到食品中，以提高食品的色泽、气味和风味。此外，他们使用食品添加剂来延长食品的储存时间。然而，一些癌症可能是由食品添加剂引起的，如许多肉里面的添加剂会与胃酸和蛋白质相互作用。因此，会产生大量的亚硝酸盐，这些亚硝酸盐可能会导致慢性生理紊乱。本章将首先阐述食品添加剂的定义，然后讨论它们的重要性和危害。最后介绍一些熟悉的食品添加剂类型，以说明为什么它们很重要。

1. 什么是食品添加剂

食品添加剂是指为改善食品品质和色、香、味，以及为防腐和加工工艺的需要而加入食品中的一类化学合成或天然物质。一些食品添加剂已被使用了几个世纪。20世纪下半叶，随着加工食品的大量出现，很多天然的和人工合成的食品添加剂引入进来。

2. 食品添加剂的用途

食品添加剂的添加量应该不对消费者健康造成风险。它具有以下几种用途：

(1) 提供食物的营养品质。

(2) 提高食品品质或稳定性，或改善其感官特性。

(3) 为有特殊饮食需求的消费者群体提供必要的食品组分。

3. 食品添加剂类型

食品添加剂按功能分类：共21类。

酸度调节剂、抗结剂、消泡剂、抗氧化剂、漂白剂、膨松剂、胶姆糖基础剂、着色剂、护色剂、乳化剂、酶制剂、增味剂、面粉处理剂、被膜剂、水分保持剂、营养强化剂、防腐剂、稳定和凝固剂、甜味剂、增稠剂、其他。

举例：

调味品：有1100~1400个天然和合成的调味品用于食品加工中。

稳定剂：这些都是用来保持食品结构稳定或使食品组织结构不变。

着色剂：90%是人造的，不含有任何营养价值。

甜味剂：旨在让食物更美味。

防腐剂：有助于保持食品新鲜，防止由真菌、酵母、霉菌和细菌引起的食品腐败。

酸/碱：为多种水果提供酸味，广泛用于碳酸软饮料。

一些常见的食品添加剂：姜黄素、柠檬黄、诱惑红、山梨酸钾、异抗坏血酸钠、瓜尔豆胶、阿斯巴甜（天冬酰苯丙氨酸甲酯）、木糖醇、糖精。

4. 食品添加剂的重要性

食品添加剂发挥了我们通常认为理应具备的功能。食品受到许多环境条件的影响，如温度变化、氧化和暴露于微生物，这些都会改变它们的原始成分。食品添加剂在维持消费者需求的食品质量和特性，保持食品安全、卫生和从农田到餐桌的吸引力方面发挥着关键作用。

5. 食品添加剂的危害

每种食品添加剂的使用范围和使用量应当符合国家标准。根据国家标准来规范使用食品添加剂是安全的，若滥用食品添加剂将对身体产生危害，甚至会致癌、致畸。

食品添加剂产生危害的原因有以下几种：

(1) 使用未经国家批准使用或禁用的添加剂品种。

(2) 添加剂使用超出规定用量。

(3) 添加剂使用超出规定范围。

（4）使用工业级代替食品级的添加剂。

食品添加剂是一把双刃剑,它有优点也有缺点。一方面,它不仅给我们带来了利益,也促进了现代食品工业的发展。另一方面,它可能被不法分子使用,从而损害我们的健康。

为了避免食品添加剂的弊端,国家应该不断完善相关法律,加强对食品生产的监管。此外,人们应该提高自身的法律和责任意识,不要只注重利益。同时,作为消费者,我们应该了解更多关于食品添加剂、食品安全的知识,并知道如何选择合适的和健康的食品。总之,我们应该尽量避免食品添加剂的缺点,充分利用其优点来提高我们对食品的兴趣。

Exercise

1. Answer questions
（1）What are food additives?
（2）What are the functions of food additives?
（3）What are the causes of the hazards of food additives?
2. Translation
（1）With the advent of processed foods in the second half of the 20th century, many more food additives have been introduced, of both natural and artificial origin.
（2）食品添加剂在维持消费者的食品质量和特性,保持消费者的需求,保持食品安全、健康,满足人们从农场到餐桌上的需求,发挥着关键的作用。

Lesson 18

Reading Material

Preservatives

Food preservatives are the additives that are used to inhibit the growth of bacteria, molds and yeasts in the food. They are the most important kind of food additives, because they play a vital role in safety of the food supply. In despite of this fact, fresh food will become inferior or unsafe when they are added chemical in order to offset the perish ability of food. Preservatives are used to prolong the shelf life of food and make sure their through that expanding period. Most importantly, they can stem bacterial degradation which will result in the production of toxins and attract food poisoning but they can also lead to some illness. The previous research shows that some modern synthetic preservatives can cause respiratory or other health problems. For example, sulfites are commonly used in wines and some dried fruits or vegetables, they may stimulate the patients with asthma.

1. Definition of preservatives

Food preservatives may be defined as substances which can kill microorganisms or prevent the growth of microorganisms. Or in simple words, preservatives are substances used in foods which have antimicrobial effects. In general, any substance that is added to foods to make it last for a longer time can be called a preservative. Some of the food additives are manufactured from the natural sources such as corn, beet and soybean, while some are artificial, man-made additives. Most people tend to eat the ready-made food available in the market, rather than preparing it at home. Such foods contain some kind of additives and preservatives, so that their quality and flavor is maintained and they are not spoiled by bacteria and yeasts. More than 3000 additives and preservatives are available in the market, which are used as antioxidants and anti-microbial agents.

2. Food preservation

Food preservation is a critical control point that influences and determines a whole range of outcomes, ranging from preservation of nutritional quality, food safety, the wholesome nature of food, texture, taste and organoleptic qualities, and consumer appeal, along with compliance to several points in the value chain that include long-term storage, long-distance transportation and marketing. In an era that is becoming increasingly global, the economics of food preservation, shipping and transportation determine not only the availability of food globally, but also the availability of food to the consumer at a reasonable price that can sustain the whole food value chain. This is especially critical in situations involving the shipping of fresh food over large distances. If the perish ability is high, the food must reach the destination in a short time to enhance the market window.

There are several factors that influence the properties of preserved food, and these factors determine the nature and method of preservation techniques that are employed. Preservation of dry foods with low water activity is relatively easy. When it comes to highly perishable foods such as meat, seafoods, fruits and vegetables, this is a challenge. In animal products, the major stress is on the prevention of microbial growth, and preservation techniques are used to achieve this goal. The shelf life of fruits and vegetables is highly variable. Rapidly respiring commodities have a very short shelf life, and consequently methods involving low temperature and anaerobic conditions are favored. This is especially true for fresh-cut fruits and vegetables. Sun drying or drying in general has been practiced as a mode of food preservation for centuries. Thus, reducing water activity is an efficient method for food preservation. The application of concentrated osmotic solutions for dehydration is another way to achieve the same goal. In every method, there is an added element of food safety and killing harmful microorganisms is essential to ensure the preservation as well as the safety of food. Food preservation methods involving thermal and non-thermal techniques have been widely employed in the food industry. At present, there are several new methods concurrently used in conjunction with traditional methods, making use of such technologies as microwaves, electricity (pulsed field), high pressure and irradiation, to provide better preservation of food.

3. Classification and introduction of preservatives

You have learnt that increasing the concentration of salt, sugar or acid in a food prevents its spoilage. Therefore, salt, sugar or acid are substances which act as preservatives.

There are two types of preservatives:

(1) Natural Preservatives: Salt, sugar, lemon juice, vinegar, oil and spices are natural preservatives.

(2) Chemical Preservatives: Potassium metabisulphate, citric acid and sodium benzoate are chemical preservatives.

For example, how does salt act as a preservative? Increasing the quantity of salt in the food changes its composition. Due to the presence of salt in the food, osmosis takes place. As a result, water comes out of the food. When there is no or less water in the food, the micro organisms are not able to grow and the food becomes safe. Salt also reduces the activity of enzymes, thus preventing the food from getting spoilt.

4. Mechanisms of preservatives

(1) Food spoilage caused by micro-organisms (three types)

①Food spoilage caused by the action of bacteria

②Food spoilage caused by the action of molds

③Food fermentation

(2) Mechanisms of food preservatives

①Destroy the structure or change the permeability of microbial cell membrane

②Interfere with microbial enzyme systems

③Other effects: cause the denaturation of microbial proteins

(3) Example: mechanisms of sorbic acid: Interfere with microbial enzymes:

①Inhibit dehydrogenases involved in fatty acid oxidation

②Inhibit sulfhydryl enzymes

Vocabulary

asthma　哮喘
degradation　降级,退化
dehydrogenase　脱氢酶
preservative　防腐剂
inferior　差的;次品
interfere　干扰,干涉,妨碍
irritant　刺激物;刺激性
mechanisms　机制,机理
offset　弥补,抵消

respiratory　呼吸道;呼吸的
shelf life　保质期
sorbic acid　山梨酸
sodium benzoate　苯甲酸钠
spoilage　腐败,变坏
stem　阻止
sulfites　亚硫酸盐
sulfhydryl　含巯基的

参考译文

防 腐 剂

食品防腐剂是用来抑制食品中细菌、霉菌和酵母生长的添加剂，它们是最重要的一种食品添加剂，因为它们在食品供应的安全问题上起着至关重要的作用。尽管如此，当新鲜食品中被添加了化学成分以延长食品的保质期的时候，它们仍然会变得劣质或不安全。防腐剂用来延长食品的保质期，并确保食品的保质期。最重要的是，它们可以阻止细菌降解，这将导致毒素的产生和引起食物中毒，进而导致疾病的发生。以往的研究表明，一些现代的合成防腐剂会引起呼吸道或其他健康问题。例如，亚硫酸盐通常用于葡萄酒和一些干果或蔬菜，它们可能刺激到哮喘病人。

1. 防腐剂的定义

食品防腐剂可以定义为可杀死微生物或防止微生物生长的物质。简言之，防腐剂是具有抗菌作用的可在食品中使用的物质。广义上，任何添加到食物中的物质使其能保存更长的时间都可称为防腐剂。这些添加剂中有些是从天然产品如玉米、甜菜和大豆中制造的，而有些是人工合成的。大多数人倾向于吃市场上现成的食物，而不是在家准备，这类食品含有某种添加剂和防腐剂，因此它们的质量和风味得以保持，不会被细菌和酵母破坏。市场上有3000多种添加剂和防腐剂，用作抗氧化剂和抗微生物剂。

2. 食品防腐

食品保藏是影响和决定一系列食品结果的关键控制点，包括保持食品营养质量、食品安全、食品的健康特性、质地、味觉和感官品质、消费者吸引力，以及符合长期存储、长途运输和营销等几个有价值的方面。在一个越来越全球化的时代，食品保鲜、运输不仅决定了全球食品的供应，而且还可以维持整个食品价值链以合理价格向消费者提供食物，这对于远距离运送新鲜食物尤其重要。如果食品易腐性高，则必须在短时间内送达目的地，以增强市场窗口。

影响食品防腐保藏特性的因素有多种，这些因素决定了所采用的储存技术和方法。保存低水分的干性食物相对比较容易，当涉及到高度易腐的食物如肉类、海鲜、水果和蔬菜，这个挑战性就比较大。对于动物产品，防腐的主要的压力在于预防微生物生长，并采用保存技术实现这一目标。水果和蔬菜的保质期是非常易变的，具有呼吸活性的商品的保质期都非常短，因此低温和厌氧的方法较受欢迎，这对于新切的水果和蔬菜尤其如此。数百年来，晒干或干燥已被实践为食品保鲜的一种模式,因此,减少水分活度是食物保存的一种有效方法,另一种能够实现同样目标的方法是浓缩渗透的脱水应用。在每一种方法中,都有一个额外的食品安全的元素,杀死有害微生物,这对于确保食物的保存和安全至关重要。食品防腐的方法包括加热技术和非加热技术的保存方法,已经广泛地应用在食品工业中。目前,有几种新方法和传统方法结合使用,比如利用微波、电(脉冲场)、高压和辐射等技术来更好的保存食物。

3. 防腐剂分类和介绍

你已经了解到,增加食物中盐、糖或酸的浓度可以防止其腐败。因此,盐、糖或酸就是起防腐作用的物质。

防腐剂有两类:

(1)天然防腐剂:盐、糖、柠檬汁、醋、油和香料都是天然防腐剂。

(2)化学防腐剂:焦亚硫酸氢钾、柠檬酸和苯甲酸钠都是化学防腐剂。

例如,盐是如何作为防腐剂?增加食物中盐的含量会改变其组成,由于食物中存在盐,会发生渗透作用。结果,水从食物中流出,当食物中没有或只有少量水时,微生物就无法生长,食物也就变得安全。盐还会降低酶的活性,从而防止食物变质。

4. 防腐的机理

(1)由微生物引起的食物腐败(三种类型)

①食物被细菌作用损坏

②食物被霉菌作用损坏

③发酵

(2)食品防腐剂的机理

①破坏结构或改变微生物细胞膜的通透性

②干扰微生物酶系统

③其他的影响:微生物蛋白失活

(3)实例:山梨酸的防腐机理:干扰微生物酶

①抑制脂肪酸氧化过程中的脱氢酶

②抑制巯基酶

Exercise

1. Answer questions

(1) What are food preservatives?

(2) What are the preservative mechanisms of food preservatives?

(3) What are the chemical preservatives?

2. Translation

(1) Food preservatives may be defined as substances which can kill microorganisms or prevent the growth of microorganisms. Or in simple words, preservatives are substances used in foods which have antimicrobial effects.

(2) At present, there are several new methods concurrently used in conjunction with traditional methods, making use of such technologies as microwaves, electricity (pulsed field), high pressure and irradiation, to provide better preservation of food.

Lesson 19

Reading Material

Food Colorants

Color is known to be one of the main factors used by consumers when they evaluate the quality and freshness of some food products. Food colorants are defined as substances capable of coloring food or improve the color of food. Food colorants are in many cases fundamental food additives because consumers judge product quality by its color. However, before a dye (natural or synthetic) is permitted for use on food, it has to be shown that it is nontoxic and noncarcinogenic. A number of natural or synthetic colorants are available and used to adjust or correct food discoloration or color change during processing or storage.

1. Why we use food colorants?

Colorant is a material that changes the color of reflected or transmitted light as the result of wavelength-selective absorption. This physical process differs from fluorescence, phosphorescence, and other forms of luminescence, in which a material emits light.

People associate certain colors with certain flavors, and the color of food can influence the perceived flavor in anything from candy to wine. Sometimes the aim is to simulate a color that is perceived by the consumer as natural, such as adding red color to glace cherries (which would otherwise be beige), but sometimes it is for effect, like the green ketchup that Heinz launched in 1999. Color additives are used in foods for many reasons including:

(1) To make food more attractive, appealing, appetizing, and informative.

(2) Offset color loss due to exposure to light, air, temperature extremes, moisture and storage conditions.

(3) Correct natural variations in color.

(4) Enhance color that occur naturally.

(5) Provide color to colorless and "fun" foods.

(6) Allow consumers to identify products on sight, like candy flavors.

2. Classification and introduction

According to the source of food colorants, they can be divided into: (1) Natural colorants. Such as curcumin, sodium copper chlorophyllin, monascus red; (2) Synthetic colorants. Such as tartrazine, brilliant blue, amaranth.

Carotenoids, chlorophyllin, anthocyanins, and betanin comprise are four main categories of plant pigments grown to color food products. Other colorants or specialized derivatives of these core groups including:

Annatto, a reddish-orange dye made from the seed of the achiote.

Caramel coloring, made from caramelized sugar.

Carmine, a red dye derived from the cochineal insect.

Lycopene.

Paprika.

Turmeric.

Blue colors are especially rare. One feasible blue dye currently in use is derived from spirulina.

Among these colorants, carotenoids are used the most, followed by red beet pigment and brown colored caramels. Yellow and red colors are used the most. Food products which are often colored are confections, beverages, dessert powders, cereals, ice creams and dairy products.

Synthetic food dyes are widely used to enhance the appearance of foods and offer certain advantages over natural colorings in that they are brighter, more stable, cheaper, and available in a greater variety of colors. However, the discovery of the toxic effects of some food dyes has led to a drastic reduction in the range of synthetic dyes that are permitted food additives in certain parts of the world. The majority of synthetic food dyes are ideal candidates for separation by CE (capillary electrophoresis). This is because they commonly contain sulfonic acid or carboxylic acid functional groups that form negatively charged colored ions at alkaline pH, thus providing a distinctive chromophore for selective detection. However, a complication in their analysis is that many food dyes are mixture of components rather than single chemical species. Although reported methods are less widespread, CE has also been applied to the analysis of natural colors such as caramels, which occur as four distinct classes according to the reactants used during sugar caramelization. Indeed, CE analysis can be used to identify and quantitate the class of caramel present in a sample.

Physical and chemical properties of some food colorants:

(1) Curcumin

①orange crystalline powder

②taste slightly bitter

③in alkaline with reddish brown, in a neutral and acid with yellow

(2) Amaranth

①dark red powder

②odorless

③can be decomposed by bacterial easily

3. Commercial abuse and health risk

Artificial colorings may be harmful. Scientists have determined a link between artificial food coloring and cancer. Other studies have linked artificial food coloring with brain tumors, ADHD (Attention Deficit Hyperactivity Disorder), and other disruptive behavior, especially in children. It's believed that the tar and hydrocarbon derivatives as well as petrochemicals used to manufacture artificial food coloring are the culprits to these diseases and disorders. None of these additives have any beneficial or nutritional value to the human body.

It is worth noting that food additives do no harm. It is not the fault of food additive, but the illegal addition. For example, in 1996, China banned food manufacturers from using Sudan I red dye to color their products due to its links to cancer and other negative health effects. However, it still discovered in 2005 that Sudan I was being used in food in many major Chinese cities. In Beijing, the Heinz Company added the red dye to chili sauce; in Guangdong, Zhejiang, Hunan, and Fujian provinces, the red dye was discovered in vegetables and noodles. Kentucky Fried Chicken (KFC) used the red dye in its 1,200 restaurants, and medicine in Shanghai also contained Sudan I.

In current situation, adding plant extracts is a natural way to add food coloring to foodstuff without all of the harmful side-effects found in artificial ones. In fact, most of these extracts are high in bioflavonoids, antioxidants, and polyphenols which are beneficial when added to your diet.

Vocabulary

carmine　胭脂红
caramel coloring　焦糖色素
dye　染料,染色
natural colorant　天然着色剂
synthetic colorant　合成着色剂
curcumin　姜黄素
sodium copper chlorophyllin　叶绿素铜钠盐
monascus red　红曲红

参考译文

食品着色剂

众所周知,颜色是消费者在评价一些食品的质量和新鲜度时使用的主要因素之一。食品着色剂被定义为能够给食物着色或改善食品颜色的物质。食品着色剂在许多情况下是基本的食品添加剂,因为消费者根据产品的颜色来判断产品的质量,然而,在允许色素(天然或合成色素)用于食品之前,必须证明它是无毒和无害的。在食品加工或储存过程中,有许多天然或合成的着色剂可供选择,用于调整或纠正食物褪色或变色。

1. 为什么要使用食品着色剂?

着色剂是由于波长选择性吸收而改变反射或透射光颜色的材料。这种物理过程不同于荧光、磷光和其他形式的发光,它们是材料在发光。

人们将某些颜色与某些风味联系在一起,食物的颜色可以影响从糖果到葡萄酒等任何东西的感知风味。有时,是为了模拟出消费者认为是天然的颜色,比如给糖渍樱桃添加红色(否则会是米色),但有时是为了效果,就像亨氏公司在1999年推出的绿色番茄酱。食品中使用着色剂的原因很多,包括:

(1)使食物更有吸引力、更吸引人、更开胃、更有涵义。
(2)抵消因暴露于光、空气、极端温度、湿度和储存条件而造成的颜色损失。
(3)校正颜色的自然变化。
(4)增强自然产生的颜色。
(5)为无色和"有趣"的食物提供颜色。
(6)允许消费者一眼就能识别出产品,如糖果口味。

2. 分类和简介

根据着色剂来源可分为:(1)天然着色剂,如姜黄素、叶绿素铜钠盐、红曲红色素;(2)合成着色剂,例如柠檬黄、亮蓝、苋菜红。

在食品着色剂中,有四种主要的植物色素:类胡萝卜素、叶绿素、花青素和甜菜素。其他着色剂或核心基团的特殊衍生物还包括:

胭脂树红,一种由胭脂树种子制成的红橙色染料;

焦糖色素,由焦糖制成;

胭脂红,一种红色染料,来源于胭脂虫;

番茄红素;

辣椒红;

姜黄;

蓝色色素特别罕见,目前使用的一种可行的蓝色染料来自螺旋藻。

在这些着色剂中,类胡萝卜素使用最多,其次是甜菜红素和焦糖色素,黄色和红色色素是最常用的。经常被着色的食品有糖果、饮料、甜点心、谷类食品、冰淇淋和乳制品。

合成色素被广泛用于增强食品的外观,并提供了优于天然色素的某些优点,因为它们更亮、更稳定、更便宜,并且有更多种颜色可供选择。然而,一些食品色素毒性效应的发现导致世界某些地区允许食品添加剂使用的合成色素数量急剧减少。大多数合成食品色素是毛细管电泳(CE)分离的理想候选对象。这是因为它们通常含有磺酸或羧酸官能团,这些官能团在碱性 pH 下形成带负电荷的有色离子,从而为选择性检测提供独特的发色团。然而,在分析中有一个复杂问题是,许多食用色素是各成分的混合物,而不是单一的化学物质。尽管报道的方法不太普遍,但是 CE 法也已经被应用于焦糖等天然色素的分析,根据焦糖化过程中使用的反应物,焦糖分为四个不同的类别。事实上,CE 分析法可以用来鉴定和定量样品中焦糖的种类。

一些食品着色剂的物理和化学特性:

(1)姜黄素

①橙色结晶粉末

②味道略带苦味

③在碱性条件下呈现红棕色,在中性和酸性条件下呈现黄色

(2)苋菜红

①深红色粉末

②无味

③可以分解

3. 商业滥用和健康风险

人造色素可能有危害。科学家已经确定了人工食用色素与癌症之间的联系。其他研究将人工食用色素与脑肿瘤、多动症(ADHD)和其他破坏性行为联

系起来,尤其是儿童。人们认为,焦油和碳氢化合物衍生物以及用于制造人造食用色素的石化产品是这些疾病和病症的罪魁祸首。这些添加剂中没有一种对人体有任何益处或营养价值。

值得注意的是,食品添加剂本身无害。这不是食品添加剂的错,而是非法添加。例如,1996年,中国禁止食品生产商使用苏丹红染料为其产品着色,因为它与癌症和其他负面健康影响有关。然而,2005年,在中国许多重要城市,仍然发现很多食品中有苏丹红。在北京,亨氏公司曾将红色染料添加到辣椒酱中;在广东、浙江、湖南和福建省,发现蔬菜和面条中有红色染料。肯德基曾在1200家餐厅中使用了这种红色染料,上海的一些药品中也含有苏丹红Ⅰ。

当前,添加植物提取物是一种在食品中添加食用色素的自然方式,而不会产生人工产生的所有有害副作用。事实上,大多数这些提取物含有丰富的生物类黄酮,抗氧化剂和多酚,这些都有益于添加到您的饮食中。

Exercise

1. Answer questions
(1) What are food colorants?
(2) What is the function of food colorants?
(3) How is the safety of food colorants?
2. Translation
(1) Synthetic food dyes are widely used to enhance the appearance of foods and offer certain advantages over natural colorings in that they are brighter, more stable, cheaper, and available in a greater variety of colors.
(2) It is worth noting that food additives do no harm. It is not the fault of food additive, but the illegal addition.

Lesson 20

Reading Material

Antioxidants

For almost half a century, food-grade antioxidants have been routinely and intentionally added to food products to delay or inhibit free radical oxidation of fats and oils and the resulting off-odours and flavour known as rancidity. Antioxidants are the chemical compounds which can delay the start or slow the rate of lipid oxidation reaction in food systems. During this time, literally hundreds of compounds—both natural and human synthetic—have been evaluated for antioxidant effectiveness and human safety.

Antioxidants are used to preserve food for a longer period of time. Antioxidants act as oxygen scavengers as the presence of oxygen in the food that helps the bacteria to grow and ultimately harms the food. In the absence of antioxidant, oxidation of unsaturated fats takes place rendering to foul smell and discoloration of food. Different kinds of antioxidants foods act in a different way, but the end result is to delay or minimize the process of oxidation in food. Some antioxidant food additives combine with oxygen to prevent oxidation and other prevent the oxygen from reacting with the food and leading to its spoilage.

The major antioxidants currently used in foods are monohydroxy or polyhydroxy phenol compounds with various ring substitutions. Hydrogen is donated to free radicals by antioxidants. Formation of a complex between the lipid radical and the antioxidant radical (free radical acceptor). The resulting antioxidant free radical does not initiate another free radical due to the stabilization of delocalization of radical electron. The resulting antioxidant free radical is not subject to rapid oxidation due to its stability.

1. Kinds of Antioxidants

According to the source of antioxidants, they can be divided into:

(1) Natural antioxidants:

①Tocopherols ($\delta > \gamma > \beta > \alpha$)

②Nordihydroguaretic Acid (NDGA)

③Sesamol

④Gossypol

(2) Synthetic antioxidants:

①Butylated Hydroxy Anisole (BHA)

②Butylated Hydroxy Toluene (BHT)

③Propyl Gallate (PG)

④Tertiary Butyl Hydroquinone (TBHQ)

The synthetic antioxidants such as BHT and BHA have antioxidant activities. However, the synthetic antioxidants have the problem with their biosafety. The best natural antioxidant sources are fruits and vegetables, as well as products derived from plants. Some good choices include blueberry, raspberry, apple, broccoli, cabbage, spinach, eggplant, and legume like red kidney beans or black bean. They're also found in green tea, black tea, red wine and dark chocolate. Usually, the presence of color indicates there is a specific antioxidant in that food.

2. Application of Antioxidants to Foods

The application of antioxidants mainly includes three aspects:

(1) Direct addition of antioxidants to oil or melted fat.

(2) Addition of antioxidants to the food after they are diluted in diluents.

(3) Spraying antioxidant solution on the food or dipping food into antioxidant solution.

In addition to the antioxidative protection contributed by individual dietary antioxidants, some researches indicate that additive and synergistic interactions occur among antioxidants. For example, synthetic antioxidants (BHA, BHT, PG) mixed with garlic extracts also exhibit remarkable antioxidative activity and synergistic effect in peanut oil. The TBHQ and tea polyphenols were effective antioxidants to almond oil. TBHQ or tea polyphenols with phytic acid or citric acid exhibited great anti-oxidation ability, but the role of Vitamin C was not significant.

Vitamins such as C, A and E can be added to foods—and they often are, such as in orange juice. One of the things those additives do is act as antioxidants in the body. There is no significant physiological difference between the added antioxidants and the ones occurring naturally in the food source. However, there's also no evidence that taking antioxidant dietary supplements work as well as the antioxidants found in food products. It's important not to overdo it on supplements because there can be too much of a good thing. With food products, it would be extremely difficult to consume an excessive amount of antioxidants.

3. Ideal Antioxidants

Although the traditional synthetic antioxidants have strong antioxidant capacity, they also have long-term potentially toxic, some even inducing teratogenic and carcinogenic effect. So, safe and high antioxidant activity natural products are more and more popular for people.

The ideal antioxidant should include the following aspects:

(1) No harmful physiological effects

(2) Not contribute an objectionable flavor, odor or color to the fat

(3) Effective in low concentration

(4) Fat-soluble

(5) No destruction during processing

(6) Readily-available

(7) Economical

(8) Not absorbable by the body

Vocabulary

antioxidant 抗氧化剂
natural antioxidant 天然抗氧化剂
synthetic antioxidant 合成抗氧化剂
synergistic interactions 协同增效作用
Butylated Hydroxy Anisole (BHA) 丁基羟基茴香醚
Butylated Hydroxy Toluene (BHT) 丁基羟基甲苯
Propyl Gallate (PG) 没食子酸丙酯
Tertiary Butyl Hydroquinone (TBHQ) 叔丁基对苯二酚
free radical 自由基

参考译文

抗 氧 化 剂

近半个世纪以来,食品级抗氧化剂常常有意或无意地被添加到食品中,以延缓或抑制脂肪和油的自由基氧化,并由此产生异味和酸败的味道。抗氧化剂是可以延缓食物体系中脂质开始氧化或减缓氧化速率的化合物。在这期间,已经评估了数百种天然和人工合成的化合物的抗氧化效果和人体安全性。

抗氧化剂被用来使食物的保存时间更长。因为食品中氧的存在有助于细菌生长,最终破坏食品,此时的抗氧化剂充当氧清除剂。在没有抗氧化剂的情况下,不饱和脂肪发生氧化,使食品变臭变色。不同种类的抗氧化剂食品以不同的方式起作用,但最终结果都是延缓或最小化食品中的氧化过程。一些抗氧化食品添加剂是与氧气结合以防止氧化,另一些抗氧化剂添加剂防止氧气与食品反应导致食品腐败。

目前在食品中允许使用的主要抗氧化剂是具有苯环取代成分的一元或多元酚类化合物。抗氧化剂将氢供给自由基,在脂质自由基和抗氧化剂自由基(自由基受体)之间形成复合物。产生的抗氧化剂自由基不会引发另一自由基,由于其稳定性,所得的抗氧化剂自由基不会被快速氧化。

1. 抗氧化剂分类

根据抗氧化剂的来源,它们可以分为:

(1)天然抗氧化剂:

①生育酚($\delta>\gamma>\beta>\alpha$)

②去甲二氢愈创木酸(NDGA)

③芝麻酚

④棉酚

(2)合成抗氧化剂:

①丁基羟基茴香醚(BHA)

②丁基羟基甲苯(BHT)

③没食子酸丙酯(PG)

④叔丁基对苯二酚(TBHQ)

人工合成的抗氧化剂如BHT和BHA等抗氧化活性较好,但其安全性受到怀疑。最好的天然抗氧化剂来源是水果和蔬菜,以及来自植物的产品。蓝莓、覆盆子、苹果、花椰菜、卷心菜、菠菜、茄子和豆类如红豆或黑豆都是比较好的选择。抗氧化剂也存在于绿茶、红茶、红酒和黑巧克力中。通常,颜色的存在表明食物中有一种特殊的抗氧化剂。

2. 抗氧化剂在食品中的应用

抗氧化剂的应用主要包括三个方面:

(1)在油或脂肪中直接添加抗氧化剂;

(2)用稀释剂稀释后,在食物中加入抗氧化剂;

(3)在食物上喷洒抗氧化剂溶液或将食物浸泡在抗氧化剂溶液中。

除了单个膳食抗氧化剂具有抗氧化作用外,一些研究表明,抗氧化剂之间会发生加成或协同增效的作用。如大蒜提取物和合成抗氧化剂(BHA、BHT、PG)混合用于花生油,都具有一定的协同抗氧化作用,尤其是与 PG 的混合使用。TBHQ 和茶多酚是对杏仁油有效的抗氧化剂,植酸、柠檬酸对茶多酚和 TBHQ 延缓杏仁油氧化都有显著的协同增效作用,而维生素 C 作用不显著。

维生素如维生素 C、维生素 A 和维生素 E 可以添加到食物中——而且通常如此,比如橙汁。这些添加剂的作用之一是在体内充当抗氧化剂。添加的抗氧化剂和食物中天然存在的抗氧化剂之间没有显著的生理差异。然而,也没有证据表明服用抗氧化剂膳食补充剂和食品中发现的抗氧化剂一样有效。重要的是不要在补充剂上吃得太多,因为可能过剩。通过食物摄入,食用过量的抗氧化剂是较困难的。

3. 理想的抗氧化剂

传统合成的抗氧化剂虽然抗氧化能力比较强,但长期食用有潜在的毒性,有的甚至会产生致畸、致癌作用,因此越来越受到人们的排斥。由于许多化学合成的抗氧化物存在着较多的安全问题,所以安全的、高抗氧化活性的天然产物就越来越受到人们的青睐。

理想的抗氧化剂应包括以下几个方面:

(1)没有有害的生理影响

(2)不会给脂肪带来令人反感的味道、气味或颜色

(3)浓度低且有效

(4)脂溶性

(5)在加工过程中不被破坏

(6)随时可用

(7)经济

(8)不被人体吸收

Exercise

1. Answer questions

(1) What is the function of food antioxidants?

(2) What are the synthetic antioxidants?

(3) How is the safety of food antioxidants?

2. Translation

(1) The major antioxidants currently used in foods are monohydroxy or

polyhydroxy phenol compounds with various ring substitutions. Hydrogen is donated to free radicals by antioxidants. Formation of a complex between the lipid radical and the antioxidants radical (free radical acceptor).

(2) The synthetic antioxidants such as BHT and BHA have antioxidants activities. However, the synthetic antioxidants have the problem with their biosafety. The best natural antioxidants sources are fruits and vegetables, as well as products derived from plants.

Lesson 21

Reading Material

Flavoring agents

Since the early days of human history, spices have been used to enhance or modify the flavor of food. Chinese has been used Cinnamon for more than 3000 years. In 1492, Columbus discovered Americas on his search for a short passage to China and India— sources of valued goods such as silk and spices.

1. Introduction of flavor

Flavorant or flavoring agents is a substance that gives another substance flavor by altering or enhancing the flavors of natural food products such as meats and vegetables, or creating flavor for food products that do not have the desired flavors such as candies and other snacks. Humans use all their five senses to analyze food quality. Food that looks and smells attractive is taken into mouth. Based on a complex sensory analysis, taste, smell, touch and hear, final decision about ingestion or rejection of food is made. Sum of those characteristics of any material taken in the mouth, perceived principally by the senses of taste and smell, and also the general pain and tactile receptors in the mouth, as received and interpreted by brain. Complex combination of the olfactory, gustatory and trigeminal sensations are perceived during tasting.

The substance of flavoring agents gives another substance a flavor, altering the characteristics of the solute, causing it to become sweet, sour, aromatic, etc. Of these chemical senses, smell is the main determinant of a food's flavor. Five basic tastes — sweet, sour, bitter, salty and umami are universally recognized, although in some cultures also include pungency, astringent, etc. The number of food smells is unbounded; a food's flavor, therefore, can be easily altered by changing its smell while keeping its taste similar. This is exemplified in artificially flavored jellies, soft drinks and candies, which, while made of bases with a similar taste, have dramatically different flavors due to the use of different scents or fragrances. The flavorings of commercially produced food products are typically created by flavorists.

2. Smell and taste

Smell is mainly perceived by the nose and olfactory epithelium and generated by volatile components. At present, the number of known volatile compounds is exceed 7,000. The large number of compounds released from food and the need to separate and identify them presents a particular challenge for the analyst, and one that demands techniques with the ability to identify the widest possible range of compounds from a single sample. In this respect, sulfur compounds often appear in the literature, primarily because of their disproportionately strong odors even at trace levels and their propensity to degrade during analysis. They are often of key importance in food aromas since they can be contributors to the distinctive aromas of certain foods although also as indicators of decay.

Taste is mostly perceived by tongue, taste buds and generated by nonvolatile constituents. Table 21.1 lists primary tastes and gustatory sensation.

Table 21.1　　　　　　　　Primary tastes

Flavor	Gustatory sensation
Sweetness	sugars
Bitterness	unpleasant, sharp, or disagreeable
Sourness	acidity(H_3O^+ ions)
Saltiness	Na^+, or other alkali ions
Umami	meatiness or relish

(1) Sweet taste

Sweet taste is predominantly elicited by carbohydrates and indicates energy-rich food sources.

(2) Bitter taste

Bitter taste is evoked by many compounds. The common denominator of most bitter compounds is their pharmacological activity or toxicity.

(3) Sour taste

Strong sour taste is also repulsive and prevents the ingestion of unripe fruits and spoiled food, which often contain acids.

(4) Salt taste

Salt taste is elicited by NaCl and contributes to electrolyte homeostasis. Consistent with this function, salt taste is attractive at low concentrations and repulsive at high concentrations.

(5) Umami taste

People taste umami through taste receptors that typically respond to glutamate, which is widely present in meat broths and fermented products and commonly added to some foods in the form of monosodium glutamate (MSG).

3. Classification of flavoring agents

Food products, whether fresh or processed, must have desirable flavors that are pleasant to the palate of the consumer. Flavorings are often added to foods to create a totally new taste, to enhance or increase the perception of flavors already present, to replace unavailable flavors, to mask less desirable flavors that are naturally present in some processed foods, or to supplement flavors already present and that had disappeared as a result of food processing.

According to the source of flavoring substances, they can be divided into:

(1) Natural flavoring substances

Obtained from plant or animal raw materials, by physical, microbiological or enzymatic processes.

(2) Nature-identical flavoring substances

Obtained by synthesis or isolated through chemical processes, which are chemically identical to flavoring substances naturally present in products intended for human consumption.

(3) Artificial flavoring substances

Not found in a natural product intended for human consumption, whether or not the product is processed.

4. Flavor formation during food processing

Flavor can be produced by thermal reactions between naturally occurring compounds in foods such as the creation of meat flavor by the thermal reactions of certain amino acids and sugars (the so-called Maillard reactions). These types of materials have been used for more than 100 years in condiment industry. Flavors generated during heating or processing by enzymatic reactions or by fermentation are generally considered to be "natural" flavors.

Other process ingredients such as hydrolyzed vegetable proteins (HVPs), autolyzed yeast extracts (AYE) and flavor enhancers may produce savory flavors.

Hydrolyzed vegetable protein produced by chemical or enzymatic hydrolysis of vegetable proteins is a typical example of a process flavor. Autolysates are produced by allowing edible yeast (i.e., brewer's yeast) to rupture. After rupture, the normally present enzymes digest the cell's proteins, carbohydrates and nucleic acids, producing flavor components.

The main reactions that lead to the formation of flavor can be listed as Maillard reactions, the Strecker degradation of amino acids, lipid oxidation and microbial and enzymatic reactions and interactions between lipids, proteins and carbohydrates.

Vocabulary

astringent　涩的
autolyze　自溶
bitter　苦的,苦味
denominator　共同特性
elicite　引出
flavoring agent　风味物质
glutamate　谷氨酸盐,味精
natural flavoring substance　天然风味物质
artificial flavoring substance　人工风味物质
olfactory epithelium　嗅觉上皮
perceive　察觉,感知
pharmacological activity　药理活性
pungency　辛辣,刺激性
predominantly　主要地
rupture　破裂
spice　调味料,香料
smell　嗅觉
taste　味觉
taste bud　味蕾
sour　酸味
umami　鲜味
volatile　易挥发的

参考译文

调 味 剂

在人类历史的早期阶段,调味料一直用于增强或改变食物的风味。中国人已经使用肉桂超过 3000 年。1492 年,哥伦布在去中国和印度寻找丝绸和香料等贵重物品的路上,发现了美洲。

1. 风味的介绍

食用香料或调味剂是指能够提供另一种物质风味的物质,它们通过改变或增强天然食品风味(如肉类和蔬菜的风味)或为不具有所需风味的食品(如糖果和其他小吃)赋予风味。人们用他们的五种感官来分析食物的质量,将看起来和闻起来都有吸引力的食物送入口中,这基于复杂的感官分析——味觉、嗅觉、触觉和听觉,最终决定该食物吃还是不吃。口腔中各种物质特征的总和,主要通过味觉和嗅觉以及口腔中的一般痛觉和触觉感受器感知,最终被大脑接收并分析,品尝过程中的感觉是嗅觉、味觉和三叉神经感觉的复杂组合。

调味剂这种物质赋予另一种物质风味,通过改变其溶质的特性,使其变得甜、酸、香等。在这些化学感觉中,气味是决定食物风味的主要因素。五种基本味觉——甜味、酸味、苦味、咸味和鲜味被普遍认可,当然有些文化中还包括辛辣、涩味等。食物的气味是无限的,因此,食物的风味很容易被改变,可以通过改变食物的气味,同时保持味道相似。这在人工调味的果冻、软饮料和糖果中有所体现,它们虽然由味道相似的基质制成,但由于使用了不同的香精或香料,风味却截然不同。市面上的调味品通常是由调味师制作的。

2. 嗅觉和味觉

嗅觉主要由鼻子和嗅觉上皮感知,是由挥发性成分产生的。目前,已知挥发性化合物的数量超过 7000 种。食物可能释放出大量的化合物,对分析员来说,分离和鉴定它们是一个特殊的挑战,也是对能够从单一样品中鉴定出尽可能多化合物的技术的挑战。在这方面,硫化合物经常出现在文献中,主要是因为它们即使含量很低,也具有异常强烈的气味,并且它们在分析过程中易于降解。它们对食物香气成分起着关键作用,因为它们可以为某些食物的独特香味做出贡献,尽管他们也是食物腐烂的标志。

味觉主要由舌头、味蕾感受,由非挥发性物质产生。

主要的口味及味觉如表 21.1 所示。

表 21.1　　　　　　　　　　主要的味觉

风味	味觉感受
甜味	糖
苦味	不愉快,尖锐
酸味	酸度(H_3O^+ 离子)
盐味	Na^+ 或其他阳离子
鲜味	肉味或鲜味

(1)甜味

甜味主要是由碳水化合物引起的,并且来自于能量丰富的食物。

(2)苦味

苦味是由许多化合物引起的。大多数苦味化合物的共同特点是它们的药理活性或毒性。

（3）酸味

强烈的酸味也是令人厌恶的,经常在未成熟的水果和变质的食物中含有酸。

（4）咸味

咸味是由 NaCl 引起的,并有助于电解质在体内平衡。与此功能一致,盐在低浓度时具有吸引力,在高浓度时具有排斥性。

（5）鲜味

人们通过味觉感受器来品尝鲜味,味觉感受器通常对谷氨酸有反应,谷氨酸广泛存在于肉汤和发酵产品中,通常以味精(MSG)的形式添加到一些食物中。

3. 风味物质分类

无论是新鲜的还是加工过的食品,都必须有令人满意的口味,让消费者感到愉悦。调味品通常被添加到食品中,以产生全新的味道,来增强或增加味道的感知,取代不可用的风味,掩盖一些加工食品中天然存在的不太理想的风味,或者补充本身含有但由于食品加工而消失的风味。

根据调味物质的来源,它们可以分为：

（1）天然调味物质

天然调味物质是指通过物理、微生物或酶促过程从植物或动物原料中获得的调味物质。

（2）天然调味物类似物

天然调味物类似物是通过化学合成或化学分离的方法获得的,其化学性质与食品中天然存在的调味物质相同。

（3）人工调味物质

无论产品是否经过加工,天然食品中未曾发现过人工调味物质。

4. 食品加工过程中风味的形成

风味可以通过食物中天然存在的化合物之间的热反应产生,例如通过某些氨基酸和糖的热反应产生肉风味(所谓的美拉德反应),这些类型的材料已应用于调味品业超过 100 年。通常认为在加热或加工过程中通过酶反应或发酵产生的香料是"天然"香料。

其他加工过程中用到的一些配料,如水解植物蛋白(HVPs)、自溶酵母提取物(AYE)和风味增强剂,可产生美好的风味。通过植物蛋白的化学或酶水解产生的水解植物蛋白是加工风味的典型例子。自溶物是通过可食用酵母(即啤酒酵母)破裂而产生的,酵母细胞破裂后,存在的酶将细胞中蛋白质、碳水化合物和核酸酶解,从而产生风味成分。

导致风味形成的主要反应包括美拉德反应、氨基酸的降解、脂质氧化、微生物和酶的反应,以及脂质、蛋白质和碳水化合物之间的相互作用。

Exercise

1. Answer questions
(1) What are flavoring agents?
(2) What are the main tastes?
(3) What flavors can be formed during food processing?
2. Translation

(1) The substance of flavoring agent gives another substance a flavor, altering the characteristics of the solute, causing it to become sweet, sour, aromatic, etc. Of these chemical senses, smell is the main determinant of a food item's flavor.

(2) Flavor can be produced by thermal reactions between naturally occurring compounds in foods, such as the creation of meat flavor by the thermal reactions of certain amino acids and sugars (the so-called Maillard reactions). These types of materials have been used by the industry for more than 100 years in savory applications. Flavors generated during heating or processing by enzymatic reactions or by fermentation are generally considered to be "natural" flavors.

Unit V
Food Quality and Safety Management
食品质量与安全管理

Lesson 22

Reading Material

Overview of Food Safety

It is universally acknowledged that the safety of food is closely related to our health. In recent years, there have seen many incidents that have made people suffer or even lose their lives; the victims also included children. Food safety is a global issue nowadays. According to the CAC (Codex Alimentarius Commission), "Food safety is the assurance that food will not cause harm to the consumer when it is prepared and/or eaten according to its intended use."

Food safety problems have caused many losses to consumers, producers and governments. Food safety problems have many causes such as economic problems, lack of technology and policy. In order to control the food hazard and food safety problems, the EU, Japan and the USA have been conducting research and have made great progress. The Chinese government has also made great efforts to regulate food safety.

The current food safety situation is optimistic on the whole, but there are still some risks: new food safety problems may appear along with technology development. It was found that most food safety problems happened at either the production stage or at the processing stage basing on several reports on food safety issues. As food can transmit pathogens which can result in the illness or death of the person or other animals. The main mediums are bacteria, viruses, mold, and fungus. It can also serve as a growth and reproductive medium for pathogens. Additionally, the reasons could be: 1) chemical fertilizer or pesticide residue; 2) veterinary residue; 3) illegal addictive; 4) bacteria exceeding the standard.

Chinese Food Safety

Food safety has been a long standing problem in China. The food safety was the most concerned issue in Chinese people, even surpassing public security, traffic safety, medical safety, etc.

In China, the major harmful factors of food include toxic animals and plants (e.g. puffer fish and toadstool), pathogenic microorganisms (e.g. Salmonella and Vibrio Parahaemolyticus) and chemical contamination (e.g. pesticide and veterinary drug residues). For example, the food safety incidents reported in 2015 in China, most were caused by chemical contamination and pathogenic microorganisms, followed by toxic animals or plants. In addition, with the rapid industrialization in China, the use of illegal additives and toxic industrial waste in food processing is a growing food safety problem, e.g. Sudan red incident in 2006, melamine scandal in 2008, gutter oil incident in 2011, etc. All the food safety problems resulted not only in public health hazards, but also in public distrust of the food industry and the government.

In consideration of the many threats to food safety in the nation, the government has made great efforts to regulate food safety. The promulgation and implementation of Food Safety Law of the People's Republic of China in 2009 marked that the reform of China's food safety laws entered a new stage. In 2010 the State Council Food Security Committee was established to improve coordination among different administrative authorities (e.g. Ministry of Health, Ministry of Agriculture, and General Administration of Quality Supervision, Inspection and Quarantine). And in 2013 China Food and Drug Administration, as the central authority, was established to improve enforcement of food safety laws and strengthen the surveillance systems. The government is in charge of legislation and supervision, but more importantly, food industries in China should take their social responsibility and put food safety ahead of economic benefits, so as to ensure food safety. Moreover, to make the public knew the truth of food safety incident, improving transparency by media and increasing communication between the government and citizens should be encouraged.

Considering the huge scale and complex situation of food industry in China, there are no shortcuts to resolving the country's food safety issues. Only through the joint efforts of the collaboration of government, food industry and consumer, China's food safety get gradual and healthy development.

Food Safety Management

The frequent food safety incidents occurring globally illustrate that food safety management in the industry is a subject that badly needs attention. With professional management of food safety, incidents and certainly their recurrence can be prevented. In the case of emerging threats, adequate management should also limit the impact of incidents.

Food safety management in the industry is not first a question of addressing food safety problems, but essentially one of taking the necessary measures to prevent them, including the necessary research and tests to confirm that the control measures are effective (validation) and properly implemented (verification). Since the introduction of the HACCP (hazard analysis and critical control point) system and other food safety management systems, the role of governments has shifted from identifying potentially unsafe food or unsafe practices to supervising and verifying the implementation of food safety management systems by industry.

The modern approach to food safety management recognizes the need for cooperation of different sectors and a role and a responsibility for each sector.

Keys to Food Safety

All to often over the last few years, shocking headlines about food safety have caught the eye of even the casual China observer. So How to make our family safer in our diet? In theory, food poisoning is 100% preventable. However, this cannot be achieved due to the number of persons involved in the supply chain, as well as the fact that pathogens can be introduced into foods no matter how many precautions are taken. The WHO (World Health Organization) suggests that there are "Five Keys to Safer Food":

(1) Prevent contaminating food with pathogens spreading from people, pets, and pests.

(2) Separate raw and cooked foods to prevent contaminating the cooked foods.

(3) Cook foods for the appropriate length of time and at the appropriate

temperature to kill pathogens.

(4) Store food at the proper temperature.

(5) Do use safe water and safe raw materials.

Vocabulary

contaminate 污染,弄脏
fertilizer 化肥
hazard 危害
implementation 实施,履行
issue 问题
legislation 立法,法律
management 管理,管理人员,管理部门
melamine 三聚氰胺
pathogen 病原体
pathogenic 致病性的
pesticide 农药,杀虫剂
promulgation 颁布
reoccurrence 复发,重现
supervision 监督,监管

参考译文

食品安全概述

众所周知,食品安全与我们的健康息息相关。近年来,发生了许多事件,使人们遭受痛苦,甚至失去生命;受害者还包括儿童。如今,食品安全已成为一个全球性问题。国际食品法典委员会(CAC)对食品安全的定义是"对食品按其原定用途进行制作、食用时不会使消费者健康受到损害的一种担保"。

食品安全问题给消费者、生产者和政府造成了许多损失。造成食品安全问题的原因很多,如经济问题、技术和政策的不足。为了控制食品危害和食品安全问题,欧盟、日本和美国一直在进行研究,并取得了很大进展。中国政府也为规范食品安全做出了巨大努力。

目前的食品安全形势总体上是乐观的,但也存在一些风险,如随着技术的发展,可能出现新的食品安全问题。根据有关食品安全问题的多份报告发现,大多数食品安全问题发生在生产阶段或加工阶段。这是由于食物可以传播病原体,导致人或其他动物生病或死亡,主要病原体就是细菌、病毒、霉菌和真菌,食物还可以作为病原体生长和繁殖的介质。另外,也可能是其他原因:(1)化肥或农药残留;(2)兽药残留;(3)非法添加物;(4)细菌超标。

中国的食品安全

食品安全问题在我国由来已久。目前,食品安全已成为国人最关心的问题,

甚至超过了公共安全、交通安全、医疗安全等。

在我国,食品中的主要有害因子包括有毒动植物(如河豚鱼和蟾蜍)、病原微生物(如沙门氏菌和副溶血性弧菌)和化学污染(如农药和兽药残留)。例如,2015年在中国报告的食品安全事件中,大多数是由化学污染物和致病微生物引起的,其次是有毒动物或植物引起的。此外,随着我国工业化进程的加快,食品加工中非法添加剂和有毒工业废弃物的使用日益成为食品安全问题,如2006年苏丹红事件、2008年三聚氰胺丑闻、2011年地沟油事件等。所有这些食品安全问题不仅对公众的健康造成危害,而且还引起了公众对食品工业和政府的不信任。

鉴于我国食品安全面临的诸多威胁,政府为规范食品安全做出了巨大努力。2009年《中华人民共和国食品安全法》的颁布实施标志着我国食品安全法改革进入了一个新的阶段。2010年,国务院食品安全委员会成立,旨在加强卫生部、农业部、质量监督检验检疫总局等不同行政部门之间的协调。2013年,中国食品药品监督管理局成立,作为中央主管部门,旨在加强食品安全法律的执行力度,加强监管体系建设。政府负责了立法和监管,但更重要的是,中国的食品行业应该承担起社会责任,把食品安全放在经济利益之上,以确保食品安全。此外,为使公众了解食品安全事件的真相,应鼓励提高媒体透明度,加强政府与公民之间的沟通。

考虑到中国食品工业规模庞大、情况复杂,解决国家食品安全问题没有捷径可走。只有通过政府、食品行业和消费者的共同努力和合作,中国的食品安全才能逐步健康发展。

食品安全管理

全球频繁发生的食品安全事件表明,该行业的食品安全管理是一个非常需要关注的问题。通过专业的食品安全管理,可以防止事故的发生,并确保其不再发生。在出现新威胁的情况下,适当的管理还应该限制事件的影响。

食品行业中的食品安全管理并不是来优先解决食品安全问题的,而是通过采取必要的措施预防食品安全问题的发生,包括一些必要的研究和测试,以确认控制措施有效并得以适当实施。自从引入HACCP(危害分析和关键控制点)体系和其他一些食品安全管理体系以来,各国政府的作用已从查明潜在的不安全食品或不安全行为转变为监督和检验食品行业执行食品安全管理体系的情况。

现代食品安全管理方法认识到不同部门合作的必要性,以及每个部门的作用和责任。

食品安全的关键

在过去的几年里,与食品安全有关的头条新闻常常刺激着中国消费者的神经。那么如何让我们在家庭的饮食上更安全呢?从理论上讲,食物中毒是可以100%预防的,然而,由于参与食品供应链的人数众多,无论采取多少预防措施,

病原体都可能进入食品,因此这一目标很难实现。世界卫生组织(WHO)建议要想获得安全的食品,有五个方面比较关键:

(1)防止因人、宠物和害虫传播的病原体污染食物。

(2)将生食和熟食分开,以免污染熟食。

(3)在适当的温度和时间内烹调食物以杀死病原体。

(4)将食物存放在合适的温度下。

(5)做到用水安全和原料安全。

Exercise

1. Answer questions

(1) What are the reasons for food safety?

(2) In daily life, how to obtain safe food as much as possible?

2. Translation

(1) Food can transmit pathogens which can result in the illness or death of the person or other animals. The main mediums are bacteria, viruses, mold, and fungus. It can also serve as a growth and reproductive medium for pathogens. Additionally, the reasons could be: 1) chemical fertilizer or pesticide residue; 2) veterinary residue; 3) illegal addictive; 4) bacteria exceeding the standard.

(2) The government is in charge of legislation and supervision, but more importantly, food industries in China should take their social responsibility and put food safety ahead of economic benefits, so as to ensure food safety. Moreover, to make the public knew the truth of food safety incident, improving transparency by media and increasing communication between the government and citizens should be encouraged.

Lesson 23

Reading Material

Food Quality Management (ISO 9000)

Quality management ensures that an organization, product or service is consistent. It has four main components: quality planning, quality assurance, quality control and quality improvement. Quality management is focused not only on product and service quality, but also on the means to achieve it. Quality management, therefore, uses quality assurance and control of processes as well as products to achieve more consistent quality.

The International Organization for Standardization (ISO) created the Quality Management System (QMS) standards in 1987. They were the ISO 9000:1987 series of standards comprising ISO 9001:1987, ISO 9002:1987 and ISO 9003:1987; which were applicable in different types of industries, based on the type of activity or process: designing, production or service delivery. The latest major revision was in 2015 and the series was called ISO 9000: 2015 series. The goal was to establish requirements for a quality management system, the implementation of which would extend to all types and business segments. The requirements of the series represented the consensus of different countries of the world. ISO 9000 series standards have been adopted by at least 110 countries around the world until 2017.

The importance of ISO 9000 is the importance of quality. Many companies offer products and services, but it is those companies who put out the best products and services efficiently that succeed. With ISO 9000, an organization can identify the root of the problem, and therefore find a solution. By improving efficiency, profit can be maximized.

In food industry, the ISO 9000 series of standards was developed to cover the requirements of both food manufacturers and auditing bodies to ensure food manufacturers met the needs of their customers while striving to meet their expectations. Many food manufacturers use ISO 9000 series for managing their business or quality systems. However, it should be noted that ISO 9000 series is a rule of general character, once adopted by the food industry, some aspects can be considered in some cases insufficient. There's not in the standard, explicit references to the risks to consumer health, the safe products, the nutritional values, the critical control points, the good manufacturing practices. Thus, some management systems for food safety also been employed to address this need.

How does ISO 9000 work?

ISO 9000 is set up as a collection of guidelines that help a company establish, maintain, and improve a quality management system. It is important to stress that ISO 9000 is not a rigid set of requirements, and that organizations have flexibility in how they implement their quality management system. This freedom allows the ISO 9000 standard to be used in a wide range of organizations, and in businesses large and small.

One important aspect of ISO 9000 is its process-oriented approach. Instead of looking at a company's departments and individual processes, ISO 9000 requires that a company look at "the big picture." How do processes interact? Can they be integrated with one another? What are the important aspects of products and services? Once this process-oriented approach is implemented, various audits can be done as a check of the effectiveness of your quality management system.

What are the ISO 9000 Principles?

The new ISO 9000: 2015 standard is based on seven quality management principles:

1. Customer Focus

The customer is the primary focus of a business. By understanding and responding to the needs of customers, an organization can correctly targeting key demographics and therefore increase revenue by delivering the products and services that the customer is looking for. With knowledge of customer needs, resources can be allocated appropriately and efficiently. Most importantly, a business's dedication will be recognized by the customer, creating customer loyalty. And customer loyalty is the best reward.

2. Leadership

A team of good leaders will establish unity and direction quickly in a business environment. Their goal is to motivate everyone working on the project, and successful leaders will minimize miscommunication within and between departments. Their role is intimately intertwined with the next ISO 9000 principle.

3. Engagement of people

The inclusion of everyone on a business team is critical to its success. Involvement of substance will lead to a personal investment in a project and in turn create motivated, committed workers. These people will tend towards innovation and creativity, and utilize their full abilities to complete a project. If people have a vested interest in performance, they will be eager to participate in the continual improvement that ISO 9000 facilitates.

4. Process approach

Understanding and managing interrelated processes as a system contributes to the organization's effectiveness and efficiency in achieving its intended results. This approach enables the organization to control the interrelationships and interdependencies among the processes of the system, so that the overall performance of the organization can be enhanced.

5. Improvement

The importance of this principle is paramount, and should a permanent objective of every organization. Through increased performance, a company can increase profits and gain an advantage over competitors. If a whole business is dedicated to continual improvement, improvement activities will be aligned, leading to faster and more efficient development. Ready for improvement and change, businesses will have the flexibility to react quickly to new opportunities.

6. Evidence-based decision making

Effective decisions are based on the analysis and interpretation of information and data. By making informed decisions, an organization will be more likely to make the right decision. As companies make this a habit, they will be able to demonstrate the effectiveness of past decisions. This will put confidence in current and future decisions.

7. Relationship management

It is important to establish a mutually beneficial relationship; such a relationship creates value for both parties. A supplier that recognizes a mutually beneficial relationship will be quick to react when a business needs to respond to customer needs or market changes. Through close contact and interaction with a supplier and customers, the organizations will be able to optimize resources and costs.

The above seven principles were chosen because they can be used to enhance corporate performance and to achieve sustained success. They formed the conceptual foundation for the ISO portfolio of quality management standards and were used to guide the development of this new standard.

Vocabulary

comprise　包含
customer　顾客
consistent　一致性
polyphenol　多酚

principle　原则
quality　质量
standard　标准

参考译文

食品质量管理(ISO 9000)

质量管理是为了确保组织、产品或服务的一致性。它有四个主要组成部分：质量计划、质量保证、质量控制和质量改进。质量管理不仅关注产品和服务质量，还关注实现质量管理的手段。因此，质量管理使用质量保证和过程控制以及产品来实现一致。

国际标准化组织(ISO)于1987年制定了质量管理体系(QMS)标准，它们是ISO 9000:1987系列标准，包括ISO 9001:1987，ISO 9002:1987和ISO 9003:1987，根据活动或过程的类型：设计、生产或服务，适用于不同类型的行业。最近一次重大修订是在2015年，称为ISO 9000:2015系列标准。ISO 9000系列标准的目标是建立质量管理体系的要求，并使其实施能延伸到所有类型和业务部门。该系列的要求代表了世界各国的共识，至2017年，全球至少110个国家已广泛采用ISO 9000系列标准。

ISO 9000的重要性在于质量的重要性。许多公司提供产品和服务，但成功的是那些高效地推出最佳产品和服务的公司。使用ISO 9000，组织可以识别问题的根源，从而找到解决方案。通过提高效率，利润可以最大化。

在食品工业中，ISO 9000系列标准的制定是为了满足食品生产商和审计机构的要求，以确保食品生产商在努力满足顾客期望的同时满足顾客的需求。许多食品制造商使用ISO 9000系列来管理他们的业务或质量体系。然而，需要注意的是，ISO 9000 系列标准是一项通用性的标准，食品工业仅仅采用 ISO 9000 系列标准是不够的。如标准中没有明确提及对消费者健康的风险、产品的安全性、营养价值、关键控制点以及良好操作规范。因此，一些食品安全管理制度也会应用起来。

ISO 9000 是如何工作的呢？

ISO 9000 是一套指导方针，帮助公司建立、维护和改进质量管理体系。必须强调的是 ISO 9000 不是一套严格的要求，组织在如何实施其质量管理体系方面具有灵活性。这种自由使 ISO 9000 标准可以广泛应用于各种组织以及大大小小的企业中。

ISO 9000 很重要的一个方面就是以过程为导向的方法。ISO 9000 要求公司关注"大局"，而不是关注公司的部门和单个流程。如流程如何交互？它们可以相互集成吗？产品和服务的重要方面是什么？一旦实施了以过程为导向的方法，就可以进行各种审核，以检查质量管理体系的有效性。

ISO 9000 的原则是什么？

新版 ISO 9000：2015 标准基于七项质量管理原则：

1. 以顾客为关注焦点

客户是企业的主要关注焦点。通过了解和响应客户需求，组织可以正确定位关键人口统计数据，并通过提供客户所需的产品和服务来增加收入。在了解客户需求的情况下，可以适当和高效地分配资源。最重要的是，企业的奉献精神将得到客户的认可，从而创造客户的忠诚度。而顾客忠诚度就是最好的回报。

2. 领导作用

有一个好的领导者团队将在商业环境中迅速团结并确定方向，他们的目标是激励每个参与项目的人，而成功的领导者将最大限度地减少部门内部和部门之间的沟通失误，他们的角色是将与下一个 ISO 9000 紧密连接起来。

3. 全员参与

将每个人都包括在业务团队中对其成功至关重要。物质的参与将导致个人对项目的投资，进而培养有积极性、有献身精神的工人。这些人将倾向于创新和创造力，并充分利用他们的能力来完成一个项目。如果人们对绩效有既定的兴趣，他们就会渴望参与 ISO 9000 推进的持续改进。

4. 过程方法

将相互关联的过程作为体系进行理解和管理，会有助于组织实现其预期结果的有效性和效率。这种方法使组织能够控制系统流程之间的相互关系和相互依赖性，从而提高组织的整体绩效。

5. 改进

这项原则至关重要，并应成为每个组织的永久目标。通过提高业绩，公司可以增加利润并获得超过竞争对手的优势。如果整个企业都致力于持续改进,改进活动将协同一致,从而实现更快、更高效的发展。随时准备改进和改变,企业将能够灵活地对新机遇作出快速反应。

6. 循证决策

有效的决策是基于对信息和数据的分析和解释。基于这种有效的决策,一个组织将更有可能做出正确的决定。公司将此作为一种习惯,他们将能够证明过去决策的有效性。这将使人们对当前和未来的决定充满信心。

7. 关系管理

建立互利的关系非常重要,这种关系为双方创造价值。当企业需要响应客户需求或市场变化时,承认互惠关系的供应商会迅速做出反应。通过与供应商和顾客的密切联系和互动,组织将能够优化资源和成本。

之所以选择上述七项原则,是因为它们可用于提高公司业绩,并持续成功。它们构成了 ISO 质量管理标准组合的概念基础,并用于指导这一新标准的发展。

Exercise

1. Answer questions

(1) What are the ISO 9000 Principles?

(2) How does ISO 9000 work?

2. Translation

(1) In food industry, the ISO 9000 series of standards was developed to cover the requirements of both food manufacturers and auditing bodies to ensure food manufacturers met the needs of their customers while striving to meet their expectations. Many food manufacturers use ISO 9000 series for managing their business or quality systems.

(2) One important aspect of ISO 9000 is its process-oriented approach. Instead of looking at a company's departments and individual processes, ISO 9000 requires that a company look at "the big picture." How do processes interact? Can they be integrated with one another? What are the important aspects of products and services? Once this process-oriented approach is implemented, various audits can be done as a check of the effectiveness of your quality management system.

Lesson 24

Reading Material

GMP and SSOP

In China, the issue of food safety has been in the public eye as never before. Foodborne disease has an enormous public health impact, as well as significant social and economic consequences. Thus, many food safety programs have been published in order to ensure safe food production and consumer protection.

Safety food programs can be set as the measures to be taken to ensure that food can be eaten without adversely affect to consumers' health. These measures aim to prevent food contamination, such as chemical contamination, physical contamination or microbiological contamination. The programs commonly used in this area are Good Manufacturing Practices (GMP), Sanitation Standard Operating Procedures (SSOP), Hazard Analysis and Critical Control Points (HACCP), and etc.

Good Manufacturing Practices (GMP)

GMP program is composed of a set of principles and rules to be adopted by the food industry in order to ensure the sanitary quality of their products. GMP contains both requirements and guidelines for manufacturing of food and drug products in a sanitary environment. These guidelines provide minimum requirements that a manufacturer must meet to assure that their products are consistently high in quality, from batch to batch, for their intended use.

GMP provides guidance for manufacturing, testing, and quality assurance in order to ensure that a manufactured product is safe for human consumption or use. Many countries have legislated that manufacturers follow GMP procedures and create their own GMP guidelines that correspond with their legislation.

The GMP requirements mainly include the installation of devices to prevent the entry of pests, contaminated water, dirt in the air, and still be designed to avoid the accumulation of dirt or physical contamination of food that is being manufactured. The equipment and the entire apparatus of materials used in industrial processing should be designed from materials that prevent the accumulation of dirt and must be innocuous to avoid the migration of undesirable particles to foods. On the production line, the procedures and steps for handling the product have to be documented, in order to ensure the standardization of safety practices. Regarding food handlers, the GMP recommend that training should be given and recycled so the concepts of hygiene and proper handling are assimilated as a working philosophy and fulfilled to the letter.

All GMP guideline follows a few basic principles:

(1) Manufacturing facilities must maintain a clean and hygienic manufacturing area.

(2) Manufacturing facilities must maintain controlled environmental conditions in order to prevent cross-contamination from adulterants and allergens that may render the product unsafe for human consumption or use.

(3) Manufacturing processes must be clearly defined and controlled. All critical processes and any changes are validated to ensure consistency and compliance with specifications.

(4) Instructions and procedures must be written in clear and unambiguous language using good documentation practices.

(5) Operators must be trained to carry out and document procedures.

(6) Records must be made, manually or electronically, during manufacture that demonstrate that all the steps required by the defined procedures and instructions were in fact taken and that the quantity and quality of the food or drug was as expected. Deviations must be investigated and documented.

(7) Records of manufacture (including distribution) that enable the complete history of a batch to be traced must be retained in a comprehensible and accessible form.

(8) Any distribution of products must minimize any risk to their quality.

(9) A system must be in place for recalling any batch from sale or supply.

(10) Complaints about marketed products must be examined, the causes of quality defects must be investigated, and appropriate measures must be taken with respect to the defective products and to prevent recurrence.

GMP has wide and effective application when all the principles cited are effectively deployed.

Sanitation Standard Operating Procedures(SSOP)

SSOP is the specific, written procedures necessary to ensure sanitary conditions in the food plant. It include written steps for cleaning and sanitizing to prevent product adulteration. Both pre-operational (before daily processing begins) and operational (during processing) sanitation needs are included in SSOP. SSOP is required in all meat and poultry processing plants. The SSOP procedures are specific to a particular plant, but may be similar to plants in the same or a similar industry. All SSOP procedures must be appropriately documented and validated.

Vocabulary

adulterant　杂物,混合物
allergen　过敏原
batch　一批
defect　缺点,缺陷
deviation　偏差
document　文件,记录
guideline　指导方针,指南
hygienic　卫生的
instruction　指令,使用说明
manufacture　制造,生产
plant　工厂
polyphenol　多酚
render　致使
retained　保留的

参考译文

GMP 与 SSOP

当前,中国的食品安全问题受到前所未有的关注。食源性疾病对公众健康产生巨大影响,并产生重大的社会和经济后果。因此,产生了很多食品安全方案用来确保食品的安全生产并保护消费者。

食品安全方案是为确保食品可以被食用而不会对消费者的健康产生不利影响的措施。这些措施旨在防止食品被污染,如化学、物理或微生物的污染。食品领域常用的方案有良好操作规范(GMP)、卫生标准操作程序(SSOP)、危害分析和关键控制点(HACCP)等。

良好操作规范(GMP)

GMP 是食品工业采用的一套确保其产品卫生和质量的制度,它包括在清洁卫生的环境中生产食品和药品的要求和指南。这些指南提供了制造商必须满足的最低要求,以确保其产品在不同批次之间的质量始终如一,满足其预期用途。

GMP 指南为生产、测试和质量保证提供指导，以确保人们在消费或使用该产品时是安全的。许多国家已经立法规定，要求生产商遵循 GMP 程序，制定符合其法规的 GMP 指南。

GMP 的要求主要包括安装防虫、防污水、防灰尘的装置，并且有避免积尘或避免食品加工过程中一些物理污染的设计。生产加工设备应使用防止污物堆积的材料进行设计，并且必须是无害的，以避免不良颗粒迁移到食品中。在生产线上，加工产品的程序和步骤必须做记录，以确保安全操作的标准化。对于食品加工者，GMP 建议给予培训并强化训练，使卫生和正确操作的理念深入脑海，进而形成一种工作理念，并不断践行。

所有 GMP 指南都遵循一些基本原则：

（1）确保生产设施处在清洁卫生的区域。

（2）确保生产设施处在受控的环境条件下，以防止杂物和过敏原造成产品的交叉污染，致使消费者使用的不安全。

（3）生产加工过程必须清晰可控。所有关键步骤和加工过程中的变化必须经过验证，以确保产品的一致性，并符合规范。

（4）说明书和操作规程文件的编写要明确无误。

（5）操作者必须经过培训，以便正确操作并做好记录。

（6）生产过程中，必须以手工或电子方式做好生产记录，以证明所有生产步骤是按规定的程序和指令要求进行的，所生产的食品或药品的数量和质量符合预期。所出现的任何偏差都应记录并做好检查。

（7）保存完整的生产记录包括分销记录，以便根据这些记录追溯各批产品的全部历史。

（8）将产品分销中影响其质量的任何风险降至最低。

（9）必须有一个从销售或供应渠道召回任何批次产品的系统。

（10）对市售产品的投诉必须进行审查，对质量缺陷的原因必须进行调查，对有缺陷的产品必须采取适当措施，防止再次发生。

当所有原则得到有效运用时，良好操作规范就能够广泛和有效的应用了。

卫生标准操作程序(SSOP)

SSOP 是指为确保食品厂卫生条件所必需的作业指导文件。它们包括用以防止产品掺杂的清洁和消毒的作业文件，涵盖了加工前和加工过程中的卫生需求。所有的肉类和家禽加工厂都需要 SSOP。特定的工厂具有特定的卫生标准操作程序，但相似的工厂 SSOP 可能类似。所有 SSOP 操作程序都要进行恰当的检查和记录。

Exercise

1. Answer questions
(1) What is GMP?
(2) What is SSOP?
2. Translation

(1) Safety food programs can be set as the measures to be taken to ensure that food can be eaten without adversely affect to the consumer's health. These measures aim to prevent food contamination, such contamination are chemical, physical or microbiological. The programs commonly used in this area are Good Manufacturing Practices (GMP), Sanitation Standard Operating Procedures (SSOP), Hazard Analysis and Critical Control Points (HACCP), etc.

(2) The GMP requirements mainly include the installation of devices to prevent the entry of pests, contaminated water, dirt in the air, and still be designed to avoid the accumulation of dirt or physical contamination of food that is being manufactured. The equipment and the entire apparatus of materials used in industrial processing should be designed from materials that prevent the accumulation of dirt and must be innocuous to avoid the migration of undesirable particles to foods.

Lesson 25

Reading Material

HACCP

Hazard Analysis and Critical Control Points (HACCP) is an internationally recognized systematic approach that is used to prevent and/or control microbial, chemical, and physical hazards within the food supply. The "farm to the fork" approach was originally designed to be used by the food processing industry to produce zero defect (no hazard) food for astronauts to consume on space flights in the USA. Within the past few decades, the overall approach has expanded to be used as the most effective method of hazard risk reduction and control in all areas of the food flow, including production, processing, distribution and retail food establishments. HACCP expanded in all realms of the food industry, going into meat, poultry, seafood, dairy, and has spread now from the farm to the fork. Nowadays, HACCP has become the universally recognized and accepted method for food safety assurance, part of food safety legislation in the EU, USA, China and many other countries.

In short, this system has a systematic and scientific approach to process control, designed to prevent the occurrence of failures, ensuring that the controls are applied in processing steps where hazards might occur or critical situations. For this, the HACCP system combines technical information updated with detailed procedures to evaluate and monitor the flow of food into an industry.

Each food production site must perform an HACCP analysis in order to secure the food safety of its products. Such an analysis must be based on the seven HACCP principles:

(1) Conduct a hazard analysis

Plan to determine the food safety hazards and identify the preventive measures the plan can apply to control these hazards. A food safety hazard is any biological, chemical, or physical property that may cause a food to be unsafe for human consumption.

(2) Identify critical control points

A critical control point (CCP) is a point, step, or procedure in a food manufacturing process at which control can be applied. As a result, a food safety hazard can be prevented, eliminated, or reduced to an acceptable level.

(3) Establish critical limits for each critical control point

A critical limit is the maximum or minimum value to which a physical, biological, or chemical hazard must be controlled at a critical control point to prevent, eliminate, or reduce that hazard to an acceptable level.

(4) Establish critical control point monitoring requirements

Monitoring is the measurement or observation that the process is operating within the critical limits (or more likely the operating limits) at the CCP. Monitoring activities are necessary to ensure that the process is under control at each critical control point.

(5) Establish corrective actions

When monitoring results show a deviation from the critical limits at a CCP, corrective action must be taken, since HACCP is designed to prevent such deviations happening in the first place. Corrective actions are intended to ensure that no product is injurious to health or otherwise adulterated as a result if the deviation enters commerce.

(6) Establish procedures for ensuring the HACCP system is working as intended

Verification ensures the HACCP plan is adequate, that is, working as intended. Verification procedures may include such activities as review of HACCP plans, CCP records, critical limits and microbial sampling and analysis.

(7) Establish record keeping procedures

The HACCP regulation requires that all plants maintain certain documents, including its hazard analysis and written HACCP plan, and records documenting the monitoring of critical control points, critical limits, verification activities, and the handling of processing deviations. All documents should be signed and dated. Appropriate documentations and records are needed to demonstrate the effectiveness of the HACCP system.

The HACCP principles outline how to establish, implement and maintain a

HACCP system. These logical steps form the basis of the by nowwell known seven principles of HACCP which are accepted internationally.

Vocabulary

document　文件
establish　建立
implementation　实现,履行
record　记录
review　检查,复审
polyphenol　多酚

参考译文

HACCP

　　危害分析与关键控制点(HACCP)是一种国际公认的用于预防和/或控制食品供应中的微生物、化学和物理危害的方法体系。这种从"农田到餐桌"的管理体系最初应用于供应美国宇航员食品的企业,用于生产零缺陷(无危害)食品。随后几十年里,这一方法已广泛运用到食品流通的各个领域,包括生产、加工、分销零售食品企业,是减少和控制危害风险的最有效方法。HACCP 体系应用领域涵盖从农田到餐桌的所有食品工业,包括肉类、家禽、海产品和乳制品。如今,HACCP 已成为欧盟、美国、中国等许多国家食品安全法规的一部分,是普遍认可和接受的一种确保食品安全的方法。

　　简而言之,HACCP 体系运用系统和科学的方法来控制过程,防止危害的发生,确保在加工过程中能够有效控制可能发生的危害或危急情况。鉴于此,HACCP 体系结合新的技术信息和详细的工作过程,以评估和监测食品的流通。

　　每种食品的生产必须进行 HACCP 分析,以确保其食品安全。这种分析必须以七项 HACCP 原则为基础:

　　(1)进行危害分析

　　进行危害分析是指做好计划,确定食品安全危害,及可用于控制这些危害的预防措施。食品安全危害是指任何可能导致食品不安全消费的生物、化学或物理的特性。

　　(2)确定关键控制点

　　关键控制点(CCP)是指食品生产加工过程中的一个点、步骤或程序,通过对

CCP 的控制，可以将预防食品危害或消除、降低到可接受的水平。

（3）建立每个关键控制点的关键限值

关键限值是指在某一关键控制点上将物理、生物、化学的危害控制到最大或最小水平，从而可防止或消除所确定的危害发生，或将其降低到可接受水平。

（4）建立关键控制点的监测措施

监测是对在已确定的 CCP 的关键限值（或是操作限值）内运行的过程进行监测或观察。监控活动对于确保过程在每个关键控制点处于受控状态是很有必要的。

（5）建立纠正措施

当某一 CCP 监控结果和关键限值有偏差时，就需要采取纠正措施，这是因为 HACCP 旨在优先防止这种偏差的发生。纠正措施的目的是确保没有对健康造成伤害或掺假的产品流入市场。

（6）建立确保 HACCP 体系正常运行的程序

验证是确保 HACCP 计划是合适的，按预期工作。验证程序可包括审查 HACCP 计划、CCP 记录、关键限值及微生物取样和分析等活动。

（7）建立记录保存程序

HACCP 法规要求所有工厂保存某些文件，包括其危害分析和书面 HACCP 计划，以及关键控制点、关键限值、验证活动和处理偏差的记录。所有文件都应签名并注明日期。适当的文件和记录是很有必要的，可表明 HACCP 体系的有效性。

以上 HACCP 原则概述了如何建立、实施和维护 HACCP 体系，这些合乎逻辑的步骤构成了国际公认的 HACCP 七项原则的基础。

Exercise

1. Answer questions

（1）What is HACCP？

（2）What are the seven basic principles of HACCP？

2. Translation

（1）HACCP expanded in all realms of the food industry, going into meat, poultry, seafood, dairy, and has spread now from the farm to the fork. Nowadays, HACCP has become the universally recognized and accepted method for food safety assurance, part of food safety legislation in the EU, USA, China and many other countries.

(2) The HACCP regulation requires that all plants maintain certain documents, including its hazard analysis and written HACCP plan, and records documenting the monitoring of critical control points, critical limits, verification activities, and the handling of processing deviations. All documents should be signed and dated. Appropriate documentations and records are needed to demonstrate the effectiveness of the HACCP system.

Lesson 26

Reading Material

ISO 22000

Food safety is linked to the presence of food-borne hazards in food at the point of consumption. Since food safety hazards can occur at any stage in the food chain, it is essential that adequate control be in place. Therefore, a combined effort of all parties through the food chain is required. The ISO (International Organization for Standardization) 22000 international standard specifies the requirements for a food safety management system.

ISO 22000 is a food safety management system that can be applied to any organization in the food chain, farm to fork. Becoming certified to ISO 22000 allows a company to show their customers that they have a food safety management system in place. This provides customer confidence in the product. This is becoming more and more important as customers demand safe food and food processors require that ingredients obtained from their suppliers to be safe.

ISO 22000 was published in 2005 with the aim of bringing together food safety and quality management systems into one auditable standard. The goal of ISO 22000 is to control, and reduce to an acceptable level, any safety hazards identified for the end products delivered to the next step of the food chain. The standard combines the following generally-recognized key elements to ensure food safety at all points of the food chain:

(1) Requirements for good manufacturing practices or prerequisite programs
(2) Requirements for HACCP principle
(3) Requirements for a management system

(4) Interactive communication between suppliers, customers, and regulatory authorities

Applicability of the ISO 22000 standard

There are many other food safety standards available, but most of them are limited in scope. ISO 22000 is the only one to cover all organizations that produce, manufacture, handle, or supply food or feed, such as:

(1) Agricultural producers

(2) Feed and food manufacturers

(3) Processors

(4) Food outlets and caterers

(5) Retailers

(6) Service providers

(7) Transportation operators

(8) Storage providers

(9) Equipment manufacturers

(10) Biochemical manufacturers

(11) Packaging material manufacturers

Highlights of the ISO 22000 standard

ISO 22000 is similar in philosophy to ISO 9001, but it contains clauses that are specific to the food industry, including:

(1) The establishment of prerequisite programs (PRPs), which define the basic conditions and activities needed to maintain a hygienic environment, both within the organization and throughout the food chain.

(2) The identification and control of food safety hazards, and the determination of an acceptable level of risk.

(3) The establishment of a HACCP plan, including the identification and monitoring of critical control points: process steps at which controls can be applied to prevent or eliminate a food safety hazard, or reduce it to an acceptable level.

(4) The handling of potentially unsafe food products to ensure that they do not enter the food chain.

(5) The establishment of a food safety team responsible for tasks such as hazard analysis, selection of control measures, establishment of PRPs, and planning of internal audits.

(6) The information and characteristics needed for both raw materials and end products to ensure that a proper hazard analysis can be conducted.

(7) The establishment of a communications plan with external parties—such as suppliers, customers, and regulatory authorities—to ensure that food safety information is available to all.

The development of ISO 22000

Prior to ISO 22000, a great number of standards had been developed in different countries, and organizations in the food sector used their own codes to audit their suppliers. The sheer number of standards (and the costs of conforming to all of them), combined with the increased globalization of the food industry, made it nearly impossible to keep up with the different requirements in the global food market. Additionally, foodborne illnesses increased significantly in all markets, resulting in both economic losses and damaged reputations.

The development of the ISO 22000 standard began in 2001, with a recommendation from the Danish Standardization Body to the secretary of ISO's technical committee. ISO then developed the standard in conjunction with the Codex Alimentarius Commission (an international body jointly established by the World Health Organization and the United Nations Food and Agriculture Organization) and experts from the food industry. In August 2005, the final draft was unanimously approved by all 23 national standard bodies participating in the working group. ISO 22000 was subsequently published on September 1, 2005.

The first major revision to ISO 22000 since its launch is expected to be published in 2018. The new version of ISO 22000 will contain a number of minor alterations that have been introduced to increase the readability and clarity of the standard, as well as some substantial changes that are more structural in nature. The revised standard will provide a new understanding of the notion of "risk". Risk is a vital concept for food businesses and the standard will distinguish between risk at the operational level (through the Hazard Analysis Critical Control Point approach, or HACCP) and risk at the strategic level of the management system (business risk) with its ability to embrace opportunities in order to reach a business's specific goals.

Vocabulary

communication　交流，沟通
prerequisite　先决条件，首要必备的
retailer　零售商
unanimously　全体一致地

参考译文

ISO 22000

食品安全与食品中的食源性危害密切相关。由于在食物链的任何阶段都可能发生食品安全风险,因此必须进行适当的控制,这就需要通过食品链上各方的共同努力。ISO 22000 国际标准的相关规定就满足了食品安全管理体系的需要。

ISO 22000 是一种食品安全管理体系,可应用于从农田到餐桌的食品链中任何组织。一个公司通过了 ISO 22000 认证,就可以在正确的位置向消费者展示他们的食品安全管理体系,这可以提高消费者对产品的信心。这种认证正变得越来越重要,因为消费者需要安全的食品,而食品加工者则需要从供应商获得的产品是安全的。

ISO 22000 于 2005 年发布,旨在将食品安全和质量管理体系整合到一个可审计的标准中,目标是将食品链中确定的安全隐患控制并降低到可接受的水平。该标准综合了以下公认的关键要素,以确保食品链各个环节的食品安全:

(1) 良好操作规范或前提方案的要求

(2) HACCP 的原则要求

(3) 对管理体系的要求

(4) 供应商、客户和监管机构之间的相互沟通

ISO 22000 标准的适用性

可供使用的其他相关食品安全标准还有很多,但大多数标准的范围有限。ISO 22000 标准是唯一涵盖了从生产、加工、处理、供应到食用的所有组织的标准,如:

(1) 农业生产者

(2) 饲料和食品制造商

(3) 加工者

(4) 食品商店和餐馆

(5) 零售商

(6) 服务供应商

(7) 运输运营商

(8) 仓储公司

（9）设备制造商

（10）生物公司

（11）包装材料生产商

ISO 22000 标准的亮点

ISO 22000 在理念上与 ISO 9001 相似,但它包含了食品行业特有的条款,包括:

（1）前提方案(PRPs)的建立,它定义了组织内和整个食品链需要保持环境卫生的基本条件和活动。

（2）食品安全危害的识别和控制,并确定可接受的风险水平。

（3）建立 HACCP 计划,包括确定和监测关键控制点,通过对关键控制点采取措施,可预防或消除食品安全危害或将其降低到可接受的水平。

（4）处理潜在不安全的食品,以确保它们不会进入食物链。

（5）建立一个食品安全小组,负责诸如危险分析、控制措施的选择、前提方案(PRPs)的建立和内部审计的规划。

（6）原材料和最终产品的信息和特性的记录,以确保进行适当的危害分析。

（7）制定与外部各方(如供应商、客户和监管机构)的沟通计划,以确保所有人都能获得食品安全信息。

ISO 22000 的发展情况

在 ISO 22000 之前,不同的国家已经制定了大量的标准,组织相关的食品部门使用他们自己的准则来审核他们的供应商。随着食品工业日益全球化,标准的数量(以及遵守所有标准的费用)几乎不可能跟上全球不同食品市场的要求。此外,所有市场的食源性疾病也显著增加,造成经济损失和声誉受损。

ISO 22000 标准的发展始于2001,最初由丹麦标准化组织向 ISO 技术委员会秘书处提出建议。ISO 随后与食品法典委员会(由世界卫生组织和联合国粮农组织联合设立的一个国际机构)和食品行业专家共同制定了该标准。2005 年 8 月,参加工作组的 23 个国家标准机构一致批准了最终草案。ISO 22000 随后于 2005 年 9 月 1 日发布。

ISO 22000 自推出以来的第一次重大修订预计将于 2018 年发布。新版的 ISO 22000 将包含一些小改动,以增加标准的可读性和清晰度,以及一些本质上更具结构性的变化。修订后的标准将对"风险"进行重新定义。风险是食品企业的一个重要概念,该标准将区分运营层面的风险(通过危害分析关键控制点方法)和管理体系战略层面的风险(业务风险),以使企业抓住机遇,实现具体的商业目标。

Exercise

1. Answer questions
(1) What is ISO 22000?
(2) What areas can ISO 22000 apply?
2. Translation

(1) ISO 22000 is a Food Safety Management System that can be applied to any organization in the food chain, farm to fork. Becoming certified to ISO 22000 allows a company to show their customers that they have a food safety management system in place. This provides customer confidence in the product. This is becoming more and more important as customers demand safe food and food processors require that ingredients obtained from their suppliers to be safe.

(2) The development of the ISO 22000 standard began in 2001, with a recommendation from the Danish Standardization Body to the secretary of ISO's technical committee. ISO then developed the standard in conjunction with the Codex Alimentarius Commission (an international body jointly established by the World Health Organization and the United Nations Food and Agriculture Organization) and experts from the food industry. In August 2005, the final draft was unanimously approved by all 23 national standard bodies participating in the working group. ISO 22000 was subsequently published on September 1, 2005.

Unit Ⅵ
Food Detection
食品检测

Lesson 27

Reading Material

Food Sample Treatment

In the analytical chemistry, during dealing with the determination of mixtures of analytes present at very low concentrations in complex matrices, every single step in the analytical workflow is relevant. From the sampling procedure up to the final data processing, every step might introduce errors compromising the quality of the final analytical result. Due to the need to assure the quality, the reproducibility and the integrity of the data, in some application areas (e. g., food chemistry and pharmaceutical analysis), international regulatory agencies require several procedures, including method validation and comprehensive statistical data treatment in order for the results to be considered acceptable. Unfortunately, in spite of the tremendous evolution of the analytical instrumentation that has occurred in recent decades, especially in chromatography and mass spectrometry (MS), complex sample analysis still cannot achieve the desired results if the samples are introduced directly into the analytical instrument without a sample pretreatment step. As a result, more extended methods have been developed to fulfill regulatory and analytical requirements, resulting in methodologies that involve several independent, complex steps.

A typical analytical workflow widely for the analysis of residue and contaminants in food matrices (e.g., veterinary drugs in meat and pesticides in fruits) employs a sample preparation step after the sampling procedure. Sample preparation aims to minimize the sample complexity and to eliminate most matrix interferences before introduction into the analytical instrument (usually a chromatograph) in order to facilitate improved isolation of the target analytes before they are introduced into the detector. Nowadays, a mass spectrometer-single (MS) or in tandem (MS/MS)-is becoming the standard detector in this application area. By following the analytical workflow described, the inclusion of a sample preparation step (usually involving an extraction technique) after the sampling step simplifies the complexity of the extract to facilitate the separation of the desired analytes from other components in the analytical column. As a result, fewer and purer analytes are introduced into the MS or MS/MS detector thus making their identification and quantification less demanding.

Even considering that all described steps are without doubt important in any analytical process, sample preparation is the step most likely to cause problems and difficulties (e.g., consumption of time, cost, contamination, poor reproducibility and low extraction yields).

In the past five decades, several extraction techniques have been utilized for sample preparation. Soxhlet (SOX) and pressurized solvent extraction techniques [e.g., supercritical fluid extraction (SFE), accelerated solvent extraction (ASE) and subcritical water extraction (SWE)] gained preference for solid sample analysis. For the sample preparation step when dealing with liquid samples, large-bore open-tubular glass liquid chromatographic (LC) columns operating at gravity pressure and liquid-liquid extraction (LLE) were the dominant techniques.

In the 1970s, SPE, a miniaturized version of the classical LC technique, was introduced and shortly became one of the preferred substitutes for LLE. Although SPE represented a considerable advance over the previously utilized techniques by lowering the amount of organic solvent utilized for sample preparation, it was far from being a "green" extraction technique due to the relatively large amounts of organic solvents still utilized. During the last decade of the twentieth century, with the development of

SPME by Pawliszyn et al. , SPME may be considered a miniaturized version of SPE. A sorption-based approach, SPME in-augurated the era of the so-called microextraction techniques.

Shortly after the success of this technique, Cramers, Sandra et al. proposed another sorption-based extraction technique, termed stir-bar sorption extraction (SBSE), a modification of the SPME technique. SBSE presented extraction yields not achieved by SPME for some analytes possessing low kow values. However, for several years, only one kind commercially available sorbent for SBSE, which limited its use to the extraction of low and slightly medium-polarity analytes.

Meanwhile, a further miniaturization of SPE, termed microextraction by packed sorbent (MEPS), was developed and commercially introduced. The conventional polymeric cartridge utilized in SPE techniques to hold the sorbent was substituted by a stainless-steel, miniaturized version termed the barrel insert and needle (BIN), which could contain any of a large number of sorbents, such as those utilized in SPE.

The main feature of these and similar techniques has been to decrease or to eliminate organic solvents during the extraction procedure, thus moving the sample preparation step towards the concept of green analytical chemistry (GAC). Several solvent-based miniaturized extraction techniques were also developed in recent decades, including dispersive liquid-liquid microextraction (DLLME), single-drop microextraction (SDME) and hollow-fiber liquid-phase microextraction (HF-LPME).

Among the several micro-techniques available nowadays for sample preparation, those based upon the use of sorptive materials are by far the most utilized. The development of new materials as sorbents in sample preparation has been widely exploited in order to achieve more selective materials with higher adsorption capacity, and to expand the availability of cheaper, more easily synthesized sorbents.

The use of micro-techniques combined with more selective sorbents allows a GAC approach based on the concept of the three Rs (replace, reduce and recycle). This approach makes possible replacement of toxic solvents with green solvents, reduced consumption of solvents and sorbents, and reuse of sorbents. Also adhering to GAC, the use of new materials in micro-techniques requires less sorbent, so they consume less solvent and precursor in the synthesis of the sorbent material.

We present a review discussing recently-developed materials for use as sorbents for sample preparation techniques in line with the GAC concept. This includes several classes of materials, the techniques in which they are utilized, and the main applications resulting from the combinations. The review focuses on synthesis, properties and advantages of the materials obtained using sol-gel techniques, ionic liquids (ILs), graphene and derived materials, restricted access materials (RAMs), immunosorbents, molecularly imprinted polymers (MIPs), molecularly imprinted monoliths (MIMs), and the micro-scale sampling techniques employing them.

The Fig. 27.1 shows a general view of the analytical steps involved in this approach. The first rectangular box (sampling) illustrates some matrices for which sample preparation can frequently be used to improve analytical results. The middle box (sample preparation) illustrates the new materials employed in the main sorbent-based techniques, surrounded by the main modern sample preparation techniques that utilize these materials. The last box (separation techniques) illustrates the separation techniques usually employed in combination (on-line or off-line) with these sample preparation techniques, concluding the typical analytical workflow utilized for the analysis of complex samples.

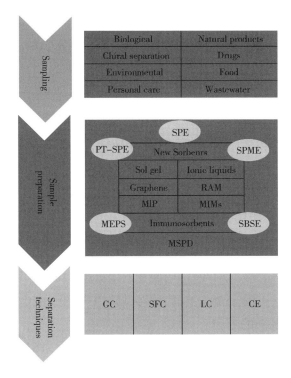

Fig. 27.1 Analytical workflow employing new materials (rectangular middle box) in the main sorbent-based sample preparation techniques.

Vocabulary

integrity　完整
pharmaceutical　药物
regulatory　监管
comprehensive　综合
chromatography　色谱分析
methodology　方法
residue　残渣
contaminant　污染物
tandem　串联
column　柱
quantification　定量
utilized　利用（过去分词）

soxhlet　索氏提取
supercritical　超临界
accelerated　加速的
subcritical　亚界
substitute　替代品
sorbent　吸附剂
conventional　传统的
cartridge　管筒
precursor　前体
synthesis　合成
immunosorbent　免疫吸附剂
rectangular　长方形

参考译文

食品样品处理

在分析化学领域，复杂基质中极低浓度的混合分析物的测定与分析工作流程的每一步都是相关的。从抽样过程到最终的数据处理，每一步都可能引入误差，影响最终分析结果的准确度。由于需要确保数据的质量、重现性和完整性，在某些应用领域（例如食品化学和药物分析），国际管理机构需要一些程序，包括检验方法验证和综合统计数据处理等以确保测定结果是可以接受的。不幸的是，尽管近几十年来分析仪器发生了巨大的变化，特别是在色谱和质谱（MS）中，如果复杂样品没有进行处理步骤就直接引入分析仪器中，样品分析仍不能达到预期的结果。因此，开发了更广泛的方法来满足监管和分析要求，从而产生了涉及多个独立、步骤复杂的方法。

在采样程序后进行样品制备步骤，是广泛运用于食品中的残留物和污染物（例如，肉中的兽药和水果中的杀虫剂）一个典型的分析工作。样品制备旨在最小化样品的复杂性，并在引入分析仪器（通常是色谱仪）之前消除大多数基质干扰，以便在引入检测器之前使目标分析物分离得更好。如今，质谱仪——单个

(MS)或串联(MS/MS)——正在成为该应用领域的标准检测器。通过遵循所述的分析工作流程,在采样步骤之后包含样品制备步骤(通常涉及萃取技术)简化了萃取物的复杂性,以促进分析柱中目标分析物与其他组分的分离。因此,将更少和更纯的分析物引入 MS 或 MS/MS 检测器,从而降低了它们的定性和定量要求。

即使考虑到在任何分析过程中所有描述的步骤都是非常重要的,样品制备仍是最有可能导致问题和困难的步骤(例如:耗时、高成本、污染、重复性差和低提取率)。

在过去的五十年中,已经有多种提取技术用于样品制备。索氏提取(SOX)和加压溶剂萃取技术[如超临界流体萃取(SFE)、加速溶剂萃取(ASE)和亚临界萃取(SWE)]优先用于固体样品分析。在液体的样品制备步骤中,在重力压力和分散液—液萃取(LLE)下操作的大口径开管式玻璃液相色谱(LC)柱是主要技术。

20 世纪 70 年代,固相萃取技术(SPE)——传统液相色谱技术的小型化版本,很快成为 LLE 的首选替代品之一。尽管降低用于样品制备的有机溶剂的量,SPE 相对先前的技术有相当大的进步,但由于仍然使用相对大量的有机溶剂,所以它远非"绿色"萃取技术。二十世纪最后十年,Pawliszyn 等发展了 SPME。SPME 可能被认为是 SPE 的小型化版本。一种基于吸附的方法,SPME 预示了所谓的微萃取技术时代。

这项技术成功后不久,克拉梅斯、桑德拉等提出了另一种基于吸附的萃取技术,称为搅拌棒吸附萃取(SBSE),对 SPME 技术进行修改。对于一些弱极性的分析物,SBSE 提取率没有 SPME 的提取率高。然而,多年来,能商业化用于 SBSE 的吸附剂仅有一种,它限制了 SBSE 技术用于萃取弱和中极性分析物质。

同时,对 SPE 技术进一步微型化,称之为微萃取(MEPS),已开发出来并商业化。在 SPE 技术中用于固定吸附剂的常规聚合物筒被小型化的不锈钢取代,称为筒或针(BIN),它可包含各种大量的吸附剂,就比如 SPE 中使用的那种吸附剂。

这些技术和类似技术的主要特点是在提取过程中减少或消除有机溶剂的使用,从而将样品制备步骤朝向绿色分析化学(GAC)发展。最近几十年来,还开发了几种基于溶剂的小型化萃取技术,包括液—液微萃取(DLLME)、单滴微萃取(SDME)和中空纤维液相微萃取(HF-LPME)等。

在当今可用于样品制备的几种微型技术中,基于使用吸附材料的那些技术目前应用最多。在样品制备过程中,吸附剂的新材料已经被广泛开发,以获得具有更高吸附容量的选择性材料,并使更便宜、更容易合成的吸附剂的使用范围扩大。

微型技术与更多选择性吸附剂的结合可以实现基于三个 R(替代,减少和回收)理念的 GAC 方法。这种方法可以用绿色溶剂替代有毒溶剂,减少溶剂和吸附剂的消耗,并可重复使用吸附剂。同样遵循 GAC 理念,在微型技术中使用新材料需要较少的吸附剂,因此它们在吸附剂材料的合成中消耗较少的溶剂和前体。

有一篇基于 GAC 理念写的综述,讨论最近开发的用作样品制备技术吸附剂的材料。这包括几类材料,使用它们的技术以及组合产生的主要应用。该文综述了利用溶胶—凝胶技术、离子液体(ILs)、石墨烯和衍生材料、限制进入材料(RAMs)、免疫吸附剂、分子印迹聚合物(MIPs)、分子印迹技术等材料的合成、性能和优点,并采用它们的微型采样技术。

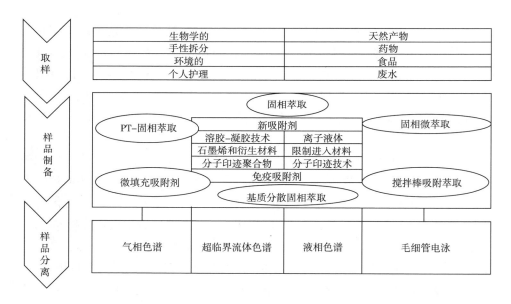

图 27.1　基于吸附剂的主要样品制备技术中采用新材料的分析工作流程

图 27.1 显示了分析方法所涉及分析步骤的一般视图。第一个矩形框(取样)列举了一些可以经常使用来改善样品制备分析结果的基质。中间的矩形框(样品制备)列举了主要吸附剂技术所使用的新材料,其周围环绕着使用这些材料的主要现代样品制备技术。最下面一个矩形框(样品分离)列举了通常与这些样品制备技术结合使用的分离技术(在线或离线),总结了用于分析复杂样品

的典型分析工作流程。

Exercise

1. Answer questions

(1) What is the purpose of the sample preparation?

(2) What are the main extraction techniques for the solid sample analysis in the past five decades?

(3) What are the disadvantages of SPE?

(4) What are the advantages of the use of micro-techniques combined with more selective sorbents?

2. Translation

(1) Sample preparation is the step most likely to cause problems and difficulties.

(2) 从抽样过程到最终的数据处理,每一步都有可能引入误差,影响最终分析结果的质量。

Lesson 28

Reading Material

Food Analysis Technologies

In recent years, consumers have become increasingly concerned about the safety and quality of their food supply. To ensure that food supplies meet the highest standards of safety and nutritional quality, robust and state-of-the-art analytical methods are essential for all food products. The rapid growth of novel raw materials and ingredients, and new processes in the food industry has also brought new challenges to food scientists. The progresses in food science and technology have driven analytical methods toward those that heavily rely on instrumentation and biochemistry, which provide higher sensitivity and accuracy. This article attempts to cover the recently developed technologies in food analysis.

1. Spectroscopy

(1) Ultraviolet and visible spectroscopy

Ultraviolet and visible spectroscopy (UV-vis) is one of the most commonly encountered techniques in food analysis. The wavelength of UV light ranges from 200~350nm, and that of Vis light ranges from 350~700nm in the spectrum. Many food components that have no or weak UV or visible absorption are analyzed with UV/Vis spectrophotometric methods after color generation through derivatization. Analyses of total phenolic, total flavonoid, total anthocyanin, and total carotenoid contents in food are good examples.

(2) Fluorimetry

During the past 20 years, the fluorescence spectroscopy is primarily considered to be a research tool in biochemistry. The use of fluorescence now has been expanded. The fluorescence spectroscopy is similar to UV-vis, but more sensitive. The instrumentation is also similar to that of the UV-vis absorption spectroscopy, except for the optical system. In a fluorometer, there is a need for two wavelength selectors, one for the emission beam and the other for the excitation beam.

(3) Infrared spectroscopy

Infrared spectroscopy refers to the measurement of the absorption of different frequencies of IR radiation by any food component in a solid, liquid, or gaseous state. IR spectroscopy can be categorized into near-IR of which the wavelength is 0.8~2.5 mm; mid-IR of which the wavelength is 2.5~15 mm; and far-IR of which the wavelength is 15~100 mm. The near-IR and mid-IR regions of the spectrum are both useful for qualitative and quantitative analysis of foods.

For food industry, the most important is attenuated total reflectance in conjunction with Fourier transform IR (FTIR) technology. It is available for the analyses of fats and oils, meats, butter, milk, even sweetened condensed milk, and juices. FTIR (Fig. 28.1) is convenient, rapid and automatable, and has dramatically simplified sample handling.

Fig. 28.1 FTIR spectroscopy Nicolet IS10

(4) Atomic absorption and emission spectroscopy

Atomic spectroscopy is widely used for accurately measuring the trace amounts of minerals in food. Although traditional chemistry titration methods for mineral analysis, such as iron, chloride, calcium, and phosphorus, remain in use today, it has been largely replaced by atomic spectroscopy. In theory, virtually all of the elements in food can be measured by atomic spectroscopy. There are two types of atomic spectroscopy, namely atomic absorption spectroscopy (AAS) and atomic emission spectroscopy (AES). AAS is based on the absorption of ultraviolet or visible radiation by free atoms in the gaseous state. Fig. 28.2 is the photograph of AAS. In contrast to AAS, the source of radiation in AES is the excited atoms or ions in the sample rather than an external source.

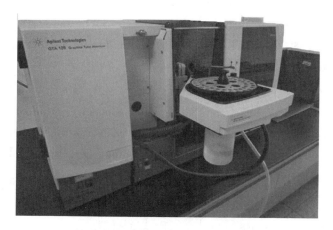

Fig. 28.2 Atomic absorption spectroscopy spectra 240FS AA

(5) Raman spectroscopy

Raman spectroscopy, a vibrational spectroscopic technique, is complementary to IR spectroscopy. It is commonly used for organic analysis. In general, the fingerprint region of organic molecules is approximately in the wavenumber range of 500~2000 nm. Another application of this technique is in studying the changes in chemical bonding, such as when a substrate is added to an enzyme. In contrast to IR, low concentrations of organic molecules in an aqueous solution are allowed to be measured due to the weak Raman scattering of water. The instrument is shown in Fig. 28.3.

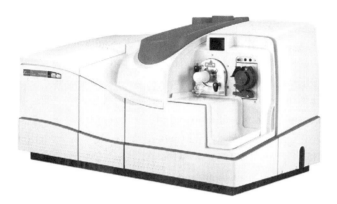

Fig. 28.3 Nexion 300 ICP-MS

2. Chromatography

(1) High performance liquid chromatography (HPLC)

Chromatography has found its use in nearly all areas of food analysis. High-or ultrahigh performance liquid chromatography compared with conventional column chromatography, HPLC is much faster. It is a convenient and widely used technology for sugar content, pesticide residues, amino acids, toxins, organic acids, lipids, vitamins, and various phytochemicals in foods. Many hydrophilic food components such as vitamin C, amino acids, phenolic compounds, and many bioactive foods are analyzed by reversed-phase HPLC, whereas both normal and reversed HPLC are used for lipophilic compounds such as vitamin E and carotenoids.

(2) Gas chromatography (GC)

GC is a separation method used to analyze thermally stable volatile substances. For example, GC has been used for the determination of fatty acids, triglycerides, flavor compounds, and many other food components, as well as pesticides, aroma compounds, and other volatile contaminants. Sample preparation is a critical step in GC analysis. It generally involves grinding, homogenization, and isolation of analytes from food samples, which may be achieved by headspace analysis, distillation, preparative chromatography, or solvent extraction.

(3) Paper and thin-layer chromatography (TLC)

Although paper chromatography is no longer widely used, TLC, because of its ease of use, relatively low cost, and greater sensitivity and reproducibility, is still used to analyze a variety of compounds in food including lipids, carbohydrates, vitamins, amino acids, and natural pigments. TLC plates with fluorescent indicators are commercially available. Compounds with UV absorption are directly detected under a UV lamp.

3. Mass spectrometry

Mass spectrometry (MS), different from the above discussed spectroscopic techniques, is based on the detection of charged molecules or fragments of a molecule. The generated ions are separated in the electrostatic field and then finally detected according to their mass-to-charge ratio (m/z). The results from ion generation, separation, fragmentation, and detection are manifested as a mass spectrum that can be interpreted to yield molecular weight or structural information of a compound. MS is now interfaced with GC and HPLC, and the hyphenated analytical methods are powerful tools and now widely used for analyzing various food components in a mixture. Structural information for the identification of unknown compounds separated and eluted from the HPLC column can be obtained using an MS detector. LC-MS has played important roles in the screening and identification of bioactive compounds such as polyphenols and carotenoids in fresh foods, and in functional foods and nutraceuticals.

4. Nuclear magnetic resonance

Nuclear magnetic resonance (NMR) spectroscopy provides important structural information for a wide variety of food components. NMR instruments (Fig. 28.4) may be configured to analyze samples in a solution or in a solid state. It may be used for the elucidation of the complete structure of complex molecules (only NMR), the 3D-imaging of fresh tissues, and the simple ingredient assays for quality assurance.

Fig. 28.4 AVANCE III 400 MHz Digital NMR spectrometer

5. X-ray method

X-ray diffraction is mainly used to identify the structure of lipids, starch, as well as proteins and enzymes. The phases of lipids can also be determined by X-ray diffraction measurements. Trace amounts of metal ions in food are detected by X-ray fluorescence, whereas X-ray adsorption spectrometry is used to determine the structural environment around the metal ions in crystalline and amorphous structures.

Vocabulary

spectroscopy 光谱学
component 组件
derivatization 衍生化
phenolic 酚类
flavonoid 类黄酮
anthocyanin 花青素
carotenoid 类胡萝卜素
fluorimetry 荧光测定法
emission 发射
excitation 激发,兴奋
infrared spectroscopy 红外光谱法
qualitative 定性
attenuated 衰减的过去式
fourier 傅里叶
automatable 自动化的
atomic absorption 原子吸收
trace 痕量
titration 滴定法

raman 拉曼
vibrational 振动
complementary 互补
bonding 成键
scattering 散射
fragment 片段
electrostatic 静电
hyphenated 连接
nuclear 原子核
magnetic 有磁性的
resonance 共振
fluorescence 荧光
crystalline 水晶般的
amorphous 无定形的
phytochemical 植物化学物质
volatile 挥发性
homogenization 均质
headspace 顶部空间

参考译文

食品分析技术

近年来,消费者越来越关注食品的质量与安全。为确保生产的食品符合最高安全和质量标准,稳定且先进的分析方法对所有食品都至关重要。新食品原料和活性成分的快速增加以及新工艺都给食品科学家带来了新的挑战。食品科技的进步推动了那些严重依赖仪器和生物化学的分析方法,这些方法具有更高的灵敏度和准确性。本文试图涵盖最近开发的食品分析技术。

1. 光谱法
(1)紫外和可见光谱法

紫外和可见光谱法(UV-vis)是食品分析中最常见的技术之一。紫外光的波长范围为200~350nm,可见光的波长范围为350~700nm。在通过衍生化反应生成颜色后,使用紫外可见分光光度法可分析许多原先没有或很弱吸收峰的食品组分。对食物中总酚、总黄酮、总花青素和总类胡萝卜素含量的分析就是很好的例子。

(2)荧光光谱法

在过去的20年中,荧光光谱法主要是生物化学中的研究工具。现在它的使用范围已经扩大了。荧光光谱法与紫外可见光谱法类似,但更敏感。除光学系统外,仪器也类似于紫外可见光谱仪。在荧光计中,需要两个波长选择器,一个用于发射光束,另一个用于激发光束。

(3)红外光谱法

红外光谱法是指通过测定固态、液态或气态的任何食物成分吸收红外光区(IR)的电磁辐射的分析方法。红外光谱可以分为近红外(波长为0.8~2.5mm);中红外(波长为2.5~15mm);远红外(波长为15~100mm)。光谱的近红外和中红外区域都可用于食品的定性和定量分析。

对于食品工业,最重要的是将衰减全反射与傅里叶变换红外(FTIR)技术相结合。它可用于分析脂肪和油、肉类、黄油、牛乳甚至加糖炼乳和果汁。FTIR(图28.1)方便、快速且可自动化,并且明显简化了样品处理。

图28.1 傅里叶红外光谱仪 Nicolet IS10

(4)原子吸收和发射光谱法

原子光谱法广泛用于精确测量食品中痕量矿物元素。虽然传统的矿物质元素分析法-滴定法,如对铁、氯、钙和磷元素测定,如今仍在使用,但它已经在很大程度上被原子光谱法所取代。理论上,食物中的几乎所有元素都可以通过原子光谱法测量。原子光谱法有两种类型,即原子吸收光谱法(AAS)和原子发射光谱法(AES)。AAS 基于气态自由原子对特定紫外或可见辐射的吸收。图 28.2 是 AAS 的照片。与 AAS 相比,AES 中的辐射源是样品中激发态的原子或离子,而不是外部源。

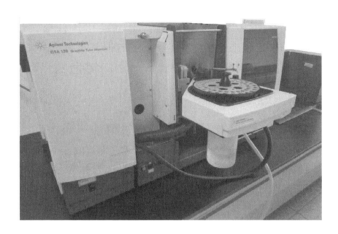

图 28.2　原子吸收光谱仪 spectra 240FS AA

(5)拉曼光谱法

拉曼光谱是一种振动光谱技术,是红外光谱的补充。它通常用于有机物分析。通常,有机物分子的指纹区域大约在 500~2000nm 的波长范围内。该技术的另一个应用是研究化学键的变化,例如当将底物加入酶时。与 IR 相反,由于水的弱拉曼散射作用,使得我们能够测量水溶液中的低浓度有机分子。仪器如图 28.3 所示。

图 28.3　拉曼光谱仪

2. 色谱法

(1)高效液相色谱法(HPLC)

色谱法几乎已应用于所有食品分析领域中。高效或超高效液相色谱与传统柱色谱相比,液相色谱更快。它是一种方便且广泛使用的技术,用于食品中的糖含量、农药残留、氨基酸、毒素、有机酸、脂类、维生素和各种植物化学物质分析。通过反相 HPLC 分析许多亲水性食物组分,例如维生素 C、氨基酸、酚类化合物和许多生物活性物质,而正向和反向 HPLC 均可用于亲脂性化合物分析,例如维生素 E 和类胡萝卜素。

(2)气相色谱法(GC)

GC 是一种用于分析易挥发性物质的分离方法。例如,GC 已用于测定脂肪酸、甘油三酯、风味物质和其他许多食品成分,以及农药、芳香化合物和其他挥发性污染物。样品制备是 GC 分析中的关键步骤。它通常涉及从食品样品中研磨,均质和分离分析物,这可以通过顶空分析、蒸馏、制备色谱或溶剂萃取来实现。

(3)纸和薄层色谱法(TLC)

虽然纸色谱不再广泛使用,但由于薄层色谱法易于使用,成本相对较低,灵敏度和重现性更高,仍用于分析食品中的各种化合物,包括脂类、碳水化合物、维生素、氨基酸和天然色素。具有荧光指示剂的薄层色谱板可商业化采购。具有紫外吸收的化合物可在紫外灯下直接检测。

3. 质谱

质谱(MS)与上述光谱技术不同,它是基于带电分子或分子片段的检测。产生的离子在静电场中分离,然后根据它们的质荷比(m/z)最终检测。离子产生、分离、碎裂和检测的图谱可以解释为产生化合物的分子质量或结构信息的质谱。MS 现在与 GC 和 HPLC 联合使用,气相-质谱和液相-质谱联用是强大的分析工具,现在广泛用于分析混合物中的各种食品成分。可以用 MS 检测器鉴定从 HPLC 柱分离和洗脱的未知化合物的结构。LC-MS 在生物活性化合物(如新鲜食品、功能性食品和营养保健品中的多酚和类胡萝卜素)的筛选和鉴定中发挥了重要作用。

4. 核磁共振

核磁共振(NMR)光谱法为食品各种成分提供重要的结构信息。NMR 仪器(图 28.4)可以配置为分析液态或固态样品。它可用于阐明复杂分子的完整结构(仅 NMR),新鲜组织的 3D 成像,以及用于保证食品质量的简单成分分析。

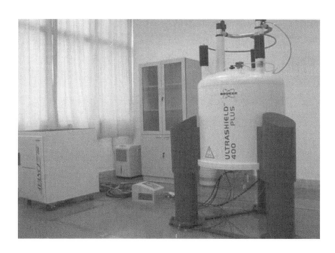

图 28.4　AVANCE III 400 MHz 数字核磁共振波谱仪

5. X 射线方法

X 射线衍射主要用于鉴定脂类、淀粉以及蛋白质和酶的结构。脂类的相位也可以通过 X 射线衍射测量来确定。可以通过 X 射线荧光来检测食品中痕量的金属离子，而 X 射线吸收光谱法用于确定结晶和无定形结构中金属离子周围的结构环境。

Exercise

1. Answer questions

（1）What is the wavelength range of UV light?

（2）What are the wavelength range of near-IR and mid-IR?

（3）How many types can atomic spectroscopy divided into?

（4）What food components are mainly analyzed by reversed phase HPLC?

2. Translation

（1）The rapid growth of novel raw materials and ingredients, and new processes in the food industry has also brought new challenges to food scientists.

（2）气相色谱法是一种用于分析易挥发性物质的分离方法。

Lesson 29

Reading Material

Food Rapid Detection Technology

The current technology is able to provide a easy, cheap, effective and safe fast microscale extraction and purification methods for various food samples. However, these methods still involve the use of sophisticated instruments such as gas chromatography – mass spectrometry (GC-MS), high performance liquid chromatography–mass spectrometry (HPLC–MS) or liquid chromatography – mass spectrometry (LC–MS) for measurement. The current demand of food contaminant detection requires a faster, on–site and preferentially naked eye detection of food sample analysis. Recent rapid testing of food contaminant is able to shorten the total detection time to 1d for microorganisms or 30min for chemicals. Although the accuracy of rapid testing is generally not as good as that of conventional testing methods, its limit of detection (LOD) is much lower than the regulated maximum residues limit (MRL) and it is competent in semi–quantitative analysis for screening purpose.

According to the government data, the number of reported food poisoning incidents in China in 2013 is 152. About 44% of the incidents were linked to pathogenic microorganisms (32.2%) or man–made chemical hazards (12.5%) whereas the remaining incidents were linked to plants and animal toxins. Bearing these statistics and the potential benefits of expanding rapid testing in mind, we subsequently review the potential contaminants that inherently exist in food and possible detection methods that could screen the substandard food items in the food supply chain.

Some of the arable lands in high population areas may encounter the problems of being over-cultivated with application of agrochemicals and/or fertilizers to increase crop yields and improve deficiency in micronutrients. Nonetheless, the overuse or misuse of these chemicals could lead to chemical and heavy metal residues in the farm produce. Environmental and artificial contaminants in food are as follows.

(1) Toxic ions in the environment

Environmental contamination by toxic ions could occur through contaminated soil and/or water. Lead contamination in water for drinking and farming is a serious problem in some areas in China. The accumulation of lead in the human body can cause kidney problems, high blood pressure, delay in physical and mental development in infants and children. The major source of contamination came from the waste discharge from smelting plants. Acceptable lead content in drinking water was restricted to 10ng/mL (GB 5749—2006). Conventional detection methods of lead involve atomic absorption spectrometry (AAS), inductively coupled plasma atomic emission spectroscopy (ICP-AES) and inductively coupled plasma mass spectrometry (ICP-MS), all of which are expensive and time-consuming. Enzyme-linked immunosorbent assay (ELISA) has also been developed but the incubation and washing steps of this method take time. Recently, a modified immunoassay test strip has been developed. It can detect lead ions with a visual LOD of 2ng/mL and no interference from other metal ions at a concentration of 1000ng/mL.

(2) Microbial contaminants

Microbial agents that contaminate agricultural productions include both Gram-negative and-positive bacteria, and microbial toxins such as bacterial, fungal and algal toxins have been implicated in numerous cases of foodborne illnesses.

Traditional plate counting of bacteria is impractical since it takes 5 ~ 7d to determine the presence of pathogenic microorganisms in food. Analytical instruments such as HPLC-MS, and GC-MS have been used to directly determine the presence of bacterial and fungal toxins. These methods have an advantage of being able to determine the analytes both qualitatively and quantitatively. GC-MS has been used to detect mycotoxins with LOD of approximately 0.1μg/kg, yet this approach is not common for commercial use as cheaper and faster alternatives such as HPLC are available. Expensive instrument and tedious preparation also hinder the more widespread use of LC-MS/MS, even though it can detect different mycotoxins simultaneously. Thin layer chromatography (TLC) with the aid of ultraviolet (UV) detector can quantitatively or semi-quantitatively determine the analytes. Its drawback is that the sample preparation and clean-up protocol is highly dependent on the type of mycotoxin. The most commonly used method to determine the presence of bacterial and fungal toxins is HPLC with UV-Vis or fluorescence detection (FD) in combination with cleanup by immunoaffinity column, which has the recovery of about 90% and LOD close to 0.05μg/kg.

Another widely used method for the detection of bacterial virulence genes in food are polymerase chain reaction (PCR) and reverse-transcription polymerase chain reaction (RT-PCR). Most probable number-PCR (MPN-PCR) shortens the detection process to within 1 d with LODs smaller than 3 or 0.3MPN/g, whereas traditional plate counting takes 3d to complete the analysis. Even though the procedures of the PCR methods take less than 3h, additional time is required for their preliminary enrichment steps, which in turn delays the whole process of detection. Moreover, the enrichment steps are bacteria-dependent and vary among different laboratories and food types. In addition, the use of DNA-DNA or DNA-RNA hybridization to determine the presence of certain virulence genes is an alternative approach for detecting bacterial toxins. Functional assays are also available for detecting bacterial and fungal toxins.

Production of mycotoxins, namely the toxic substances produced by fungi, is a worldwide common problem in crops. Mycotoxins cripple human health by inhibiting protein synthesis, damaging macrophage systems and weakening lung functions. Mycotoxin poisoning causes different symptoms depending on the type of mycotoxin, the dosage and the duration of exposure, as well as the body condition of the exposed person. These microbial toxins pose huge threat on food safety and human health. Early detection of the causative agents of foodborne illnesses is one of the best strategies to prevent foodborne outbreaks. As the symptoms of foodborne illness can be developed as quickly as within a few hours after consumption of contaminated food, establishment of fast, accurate and easy-to-use tests for on-site detection of these toxins or in general examination facilities is very important. Immunological assays are also commonly used to detect food toxicants. Due to the low molecular weight (<1ku) of many reported mycotoxins, they are usually detected by using competitive immunoassays with ELISA configuration. Fluorescence polarization immunoassay (FPIA) has been further developed to detect mycotoxins. This method does not require any additional manipulation as in the ELISA format and has LOD as low as 0.3~30ng/mL for several major mycotoxins. Despite their practical use, these assays are less useful than lateral flow devices (LFD) or commonly known as test strips, for which do not require sequential addition of reagents.

Utilizing magnetic nanoparticles (MNPs) in immunoassay is a relatively new strategy to enhance the performance of traditional ELISA assay. MNPs can be easily separated or enriched by applying a magnetic field. Researchers take advantage of this phenomenon to create MNPs-antibody conjugate, together with modified AuNPs for simultaneous detection of bacterial toxins. There is an urgent need to develop not merely rapid, but sensitive and accurate test methods for food contaminants and toxicants for in situ detection in China. Test strips based on ELISA in combination with LFD have the advantages of being efficient and easy to use for on-site testing of food served or in the food industry, allowing the monitoring of the quality of raw food materials at the early stages of food production. Unfortunately, it is impossible to use merely one test method to detect all possible microbial toxins. Moreover, there are no low-cost and convenient tests commercially available for the public. What can be done, however, is to apply the specific test methods for the natural foods that may contain certain toxins in the food industry.

(3) Pesticide

Today, an average of 15% of the crops worldwide is lost due to pests. This urges the development and applications of pesticide to meet the growing population. However, this leads to the problem of food pesticide residues which are neurotoxic. Standard pesticide analysis methods depend mainly on GC and HPLC which are time-consuming, with complex sample preparation procedures, expensive and only allow off-site analysis. Therefore, rapid tests that can act as rapid, straightforward, simple operating, cheap and naked eye detectable screening tools are being developed for the high through-put screening of major classes of pesticides.

The development of immunoassay on pesticide has become mature enough to give reliable screening within 10min. With the increasing demand for more efficient and more economical tools for screening of pesticides, multi-analyte rapid testing that detects two or more pesticides simultaneously is one of the rational solutions. There are two approaches in multi-analyte testing. The first approach is to combine two or more single-analyte assays onto one testing strip. The challenge of this approach is the cross interference of the target analytes. Nevertheless, this strip is able to give compatible quantification results to GC-MS and HPLC-FD on green pepper, red pepper and tomato matrices with 10 and 200ng/mL visual LODs for atrazine and carbaryl respectively. In contrast to the single-analyte approach, the second multi-analyte testing approach is to detect a class of pesticides by one antibody or a specific reaction mechanism that reacts with that class of pesticides. This approach aims to screen particular sample with the maximum limit of a class of pesticide which is the total quantity of pesticides. Further analysis is required only for those samples that exceed the total maximum limit. For instance, the major classes of pesticides include organophosphate (OP) pesticides and carbamate (CM) pesticides.

(4) Pest-resistant genetically modified (GM) gene

Despite the use of pesticide, the use of transgenic expression in several crops has been proved to be an efficient protection control against pests. On the one hand, pest resistance transgenic expression is characterized in highly specific for certain hosts, non-toxic to human and low environmental persistence. On the other hand, concerning about the unpredictable effects of the gene transfer or gene insertion, the attitudes of consumers towards GM foods remain ambivalent. More than 40 countries have adopted either voluntary or mandatory food labeling policies. The screening assay of GMO relies on multiplex PCR which requires multi-set of primers targeting different genes and a long amplifying cycle to detect GMO in more than an hour. Rapid detection tests for GM food is thus in a huge demand.

In contrast to the PCR method, test strip can identify the presence or absence of gene in less than 10min. For example, a colloidal gold based immuno-chromatographic strip has been developed to detect vegetative insecticidal protein (Vip), which is used for transgenic expression in several crops, with a LOD at 100 ng/mL for the detection of Vip. This colloidal gold based test strip is able to give a positive result of Vip on GM brinjal leaf while a negative result on non-GM brinjal leaf.

(5) Other rapid testing methods for pesticide

In addition to test strip, surface-enhanced raman spectroscopy (SERS) has also been applied for the rapid detection of OP pesticides. SERS is an analytical technique utilizes signal enhancement phenomenon on the surface of noble metals and transition metals. Today, portable SERS devices have been developed and SERS has become one of the potential methods for on-site detection of pesticide residues on food.

Conventional methods involving HPLC, GC, MS or ICP are limited by their sizes, gaseous supply and power supply for in-situ determination of analyte. Today, hand-held MS has demonstrated successful application in in-situ detection of thiabendazole and diphenylamine on orange and apple surfaces, respectively. Although the detection limit is about 15mg/kg of diphenylamine on apple surface which is slightly higher than the Chinese National Standard pesticide residue limit (GB 2763—2014, 5 mg/kg). This example shows an alternative way of the future development on rapid food testing.

To date, there is no approved test kit for the rapid screening of pesticide residues in the EU or the US, although many scientists have already demonstrated that the competence of test strips are comparable to those of the conventional testing methods.

Vocabulary

rugged　坚固的
sophisticated　复杂的
naked　裸露
screening　筛选
poisoning　中毒
artificial　人工
arable　耕地
lead　铅
kidney　肾脏
smelting　冶炼
immunoassay　免疫测定
algal　藻
implicated　有牵连的
mycotoxin　真菌毒素
tedious　单调乏味的
TLC　薄层色谱
protocol　协议
immunoaffinity column　免疫亲和柱
virulence　毒性
probable　可能的
hybridization　杂交
cripple　削弱
macrophage　巨噬细胞
symptom　症状
magnetic nanoparticles　磁性纳米颗粒

neurotoxic　神经毒性
straightforward　简单的
strip-based immunoassay　试纸免疫测定
thiabendazole　噻苯咪唑
neurological　神经上的
compatible　兼容
pepper　胡椒
atrazine　阿特拉津(一种除草剂名)
carbaryl　胺甲萘
organophosphate　有机磷
carbamate　氨基甲酸酯
transgenic　转基因
ambivalent　矛盾的
voluntary　自愿的
mandatory　强制性的
primers　引物
colloidal　胶体
vegetative insecticidal protein　营养期杀虫蛋白
brinjal leaf　茄子叶
SERS　表面增强拉曼光谱法
noble　贵族
transition　过渡
portable　便携式
diphenylamine　二苯胺

参考译文

食品快速检测技术

目前的技术能够为各种食品样品提供简单、廉价、有效和安全的快速微量提取和纯化方法。但是,这些方法仍然涉及使用诸如气相色谱-质谱(GC-MS)、高效液相色谱-质谱(HPLC-MS)或液相色谱-质谱(LC-MS)等精密仪器来进行测定。目前要求对食品污染物的检测是对食品样品能优先肉眼检测和快速现场分析。当前对食品污染物的快速检测,能够将微生物的总检测时间缩短至1d,对化学物质的检测时间缩短至30min。尽管快速检测的准确性一般不如传统检测方法,但其检测限(LOD)远低于规定的最大残留限量(MRL),并且它在半定量分析中具有筛选的目的。

根据政府部门统计数据,2013年中国报告的食物中毒事件数量为152起。约44%的事件与病原微生物(32.2%)或人造化学物质危害(12.5%)有关,而其余事件与植物和动物毒素有关。考虑到这一统计数据以及扩大快速检测的潜在益处,我们随后概述了食品中存在的潜在污染物,以及可能检测食品供应链中不合格食品的检测方法。

在人口密集地区,一些耕地可能会遇到过度耕种的问题,为增加作物产量和改善微量营养物质缺乏而使用农药和/或化肥。尽管如此,过度使用或滥用这些化学品可能会导致农产品中的化学物质和重金属残留。食物中的环境和人造污染物如下所示。

(1)环境中的有毒离子

被污染的土壤和/或水可能造成环境污染。在中国的一些地区,饮用水和农业的铅污染是一个严重的问题。铅在人体内的蓄积会导致肾脏问题、高血压以及延缓婴幼儿身心发育。主要的污染源来自冶炼厂的废物排放。饮用水中可接受的铅含量限制在10ng/mL(GB 5749—2006)。铅的常规检测方法涉及原子吸收光谱法(AAS)、电感耦合等离子体原子发射光谱法(ICP-AES)和电感耦合等离子体质谱法(ICP-MS),所有这些都成本高且耗时。酶联免疫吸附试验法(ELISA)也已开发出来,但该方法的孵育和洗涤步骤需要时间。最近,已经开发出了改良的免疫测定试条。它可以检测铅离子,其目测LOD为2ng/mL,浓度为1000ng/mL时不受其他金属离子的干扰。

(2)微生物污染物

污染农业生产的微生物包括革兰氏阴性和革兰氏阳性细菌,微生物毒素如细菌、真菌和藻类毒素,它们与许多食源性疾病有关。

传统的平板计数法是不切实际的,因为它需要 5~7d 来确定食物中病原微生物的存在。诸如 HPLC-MS 和 GC-MS 等分析仪器已被用于直接确定细菌和真菌毒素的存在。这些方法的优点是既能定性又能定量地测定分析物。GC-MS 已用于检测真菌毒素,LOD 大约 $0.1\mu g/kg$,但这种方法在商业用途中并不常见,因为可以使用更便宜和更快的替代方法,如 HPLC。昂贵的仪器和繁琐的制备工作也阻碍了 LC-MS/MS 更广泛的应用,尽管它可以同时检测出不同的霉菌毒素。借助紫外(UV)检测器的薄层色谱(TLC)可定量或半定量测定分析物。其缺点是样品制备和清洁方案高度依赖于霉菌毒素的类型。确定细菌和真菌毒素存在的最常用的方法是使用配有紫外检测器(UV-Vis)或荧光检测器(FD)的 HPLC 结合免疫亲和柱清洁,其具有约 90%的回收率,LOD 接近 $0.05\mu g/kg$。

另一种广泛使用的检测食物中细菌毒力基因的方法是聚合酶链式反应(PCR)和逆转录聚合酶链式反应(RT-PCR)。最可能的数字 PCR(MPN-PCR)将检测过程缩短至 1d 内,LOD 小于 3 或 $0.3MPN/g$,而传统平板计数则需要 3d 才能完成分析。即使 PCR 方法的操作时间少于 3h,但其预先浓缩步骤需要额外的时间,这又延长了整个检测过程。而且,富集步骤依赖于细菌,并且在不同的实验室和食物类型中有所不同。另外,使用 DNA-DNA 或 DNA-RNA 杂交来确定某些毒力基因的存在是检测细菌毒素的另一种方法。功能分析也可用于检测细菌和真菌毒素。

真菌毒素的产生是世界范围内普遍存在的农作物问题。真菌毒素通过抑制蛋白质合成,破坏巨噬细胞系统并削弱肺功能而削弱人体健康。真菌毒素中毒根据霉菌毒素的种类,暴露的剂量和持续时间以及接触者的身体状况而导致不同的症状。这些微生物毒素对食品安全和人体健康构成巨大威胁。早期发现食源性疾病的致病因素是预防食源性疾病暴发的最佳策略之一。由于食源性疾病的症状可以在食用受污染食品后的几个小时内迅速发展,所以建立快速、准确和易于现场使用的毒素检测方法或一般检查设备是非常重要的。免疫学分析也常用来检测食物中的毒物。由于许多报道的真菌毒素均为低分子质量蛋白质(<1ku),通常通过使用 ELISA 的竞争性免疫测定法来检测它们。为检测真菌毒素,荧光偏振免疫分析(FPIA)又得到了进一步发展。该方法不需要像 ELISA 那样进行任何额外的操作,并且对于几种主要的真菌毒素,LOD 低至 $0.3~30ng/mL$。尽管有实际用途,但这些测定法不如横向流动装置(LFD)或通常称为测试条的有用,因为它们不需要连续添加试剂。

利用磁性纳米颗粒(MNPs)进行免疫测定是一种相对较新的提高传统 ELISA 测定性能的策略。通过施加磁场可以很容易地分离或富集 MNP。研究人员利用这一现象创建 MNPs-抗体偶联物,与修饰的 AuNPs 一起同时用于检测细菌毒素。中国迫切需要开发出对食品污染物和毒物进行原位检测的快速、灵敏和准确的检测方法。由于基于 ELISA 与 LFD 结合的试纸具有高效且易于使用的优点,可以用于食品或食品工业的现场测试,因此它被允许用在食品生产的早期阶段以监测原料质量。不幸的是,仅仅使用一种测试方法来检测所有可能的微生物毒素是不可能的。此外,市面上也没有廉价且方便的测试可供公众使用。然而,我们可以做的是对食品工业中可能含有某些毒素的天然食品应用特定的测试方法。

(3) 农药

今天,全球平均有 15% 的作物因害虫而损失。这促使农药的开发和应用满足日益增长的人口需求。但是,这也导致了具有神经毒性的食物农药残留问题。标准农药分析方法主要依赖于 GC 和 HPLC,这些方法既费时,又有复杂的样品制备程序,价格昂贵且只能进行非现场分析。因此,正在开发基于主要类别农药的快速、直接、操作简便、价格便宜且能够肉眼检测的高通量筛查的快速测定方法。

农药免疫分析的发展已经足够成熟,可以在 10min 内进行可靠的筛选。随着对农药筛选更高效和更经济的需求不断增加,同时检测两种或更多种农药的多分析物快速检测是合理的解决方案之一。在多分析物测试中有两种方法。第一种方法是将两种或更多种单分析物测定结合到一个测试条上。这种方法的挑战是目标分析物的交叉干扰。尽管如此,该试纸还是可以在青椒、红辣椒和番茄基质上通过 GC-MS 和 HPLC-FD 分别得到阿特拉津(农药)和胺甲萘(农药)相容的定量结果,肉眼观察 LOD 分别是 10ng/mL 和 200ng/mL。与单分析物方法相反,第二种多分析物测试方法是通过一种抗体或与该类农药发生反应的特定反应机制来检测一类农药。这种方法的目的是筛选特定的样本,其中农药的最大限制是农药总量。只有超过最大限量的样本才需要进一步分析。例如,农药的主要类别包括有机磷(OP)农药和氨基甲酸酯(CM)农药。

(4) 抗虫基因修饰(GM)基因

尽管使用了杀虫剂,但是在几种作物中使用转基因已被证明是有效防治害虫的控制措施。一方面,抗虫转基因表达的特点是对某些宿主具有高度特异性,对人体无毒,环境持久性低。另一方面,考虑到基因转移或基因插入的不可预见性,消费者对转基因食品的态度仍然是矛盾的。40 多个国家采用了自愿或强制性食品标签政策。GMO 的筛选测定依赖于多重 PCR,其需要一个多小时针对不同基因的设计多组引物和长时间扩增以检测 GMO。因此转基因食品的快速检测是一个巨大的需求。

与 PCR 方法相比,测试条可以在 10min 内识别基因的存在与否。如基于胶体金的免疫层析试纸条检测几种作物中进行转基因表达的营养型杀虫蛋白

(Vip)，检测限为100ng/mL，这种胶体金测试条能够在转基因茄子叶上产生阳性结果，而在非转基因茄子叶上产生阴性结果。

(5) 其他农药快速测试方法

除了试纸条之外，表面增强拉曼光谱仪(SERS)也被用于快速检测有机磷农药。SERS是一种利用贵金属和过渡金属表面上的信号增强现象的分析技术。如今，便携式SERS设备已经开发出来，SERS已成为现场检测食品上农药残留的潜在方法之一。

涉及HPLC、GC、MS或ICP的常规方法受到仪器尺寸、气体供应和电源只能原位测定分析物的限制。今天，手持MS已经证明分别在橙子和苹果表面上分别原位检测噻苯咪唑和二苯胺杀虫剂的成功应用。尽管苹果表面二苯胺杀虫剂的检测限值约为15mg/kg，略高于中华人民共和国国家标准农药残留限量(GB 2763—2014，5mg/kg)，但该实例显示了未来食品快速检测发展的另一种方式。

迄今为止，虽然许多科学家已经证明测试条的能力可与传统测试方法的能力相媲美，但在欧盟或美国仍没有被批准能上市的快速筛查农药残留的试剂盒。

Exercise

1. Answer questions

(1) What are the conventional detection methods of lead?

(2) Why do we need rapid detection method to detect the presence of pathogenic microorganisms in food?

(3) What are the standard pesticide analysis methods depend on?

(4) What is the SERS?

2. Translation

(1) Nonetheless, the overuse or misuse of these chemicals could lead to chemical and heavy metal residues in the farm produce.

(2) 今天，全球平均有15%的作物因害虫而损失。

Lesson 30

Reading Material

Detection of Food Microorganisms

The use of microorganisms in food production dates back to many centuries and predates even the discovery of the microbial world. Despite an ignorance of the existence of microorganisms, a number of fermented foods became a part of the diet of different cultures produced using a recipe but not a knowledge of the underlying agents that caused the transformation.

Technology for the detection and characterization of microorganisms is about 100 years old and occurred on two fronts. The first was the visualization of the microbial world which has largely been attributed to the work of Antonie Van Leeuwenhoek whose handmade magnifying lenses allowed him to see things on the order of a few microns. The second was the discovery of methods for culturing organisms which was primarily the work of Robert Koch in the 1880s. The latter development allowed some individual microorganisms to be isolated and grown in the laboratory. Since that time, a vast number of different growth medias have been formulated, and along with specific culture conditions, a wide array of microorganisms can now be cultured from foods. There is, however, no universally successful method for culturing microorganisms from foods, and some prove illusive (e. g. , Campylobacter) leading to a vast underestimation of the number of culture-confirmed food-borne illnesses.

For approximately 75 years, microorganisms were detected through a culture-based approach, and their characterization as well as the entire taxonomy has been mainly based upon microscopic analysis and biochemical tests. Many of these techniques have been adapted from the clinical diagnostics field where samples are usually more uniform and less complex than foods. Nucleic acids were discovered in

1940s by Avery, MacLeod and McCarty, but the use of nucleic acids as routine diagnostic tools would wait for the technology to advance to a point where the methods were practical to implement. Two developments changed the landscape of microbial diagnostics—nucleic acid sequencing and polymerase chain reaction (PCR). Both evolved out of the field of recombinant DNA technology and the establishment of methods for the in vitro manipulation of nucleic acids coupled to the ability to transform a host organism with these recombinant nucleic acid molecules.

Nucleic acid sequence analysis has evolved over the past 30 years from a very manual, laborious method that required days to complete even very short sequences to an automated, instrument – driven process that can complete an entire microbial genome in a matter of hours. The transition from a manual process to an automated one was a result of an interdisciplinary effort exploiting biologists and engineers to incorporate the basic enzyme – based nucleic acid sequence analysis into an instrument. The first automated DNA sequencers were introduced in the 1980s by Applied Biosystems, and a technological revolution quickly ensued resulted in exponential increases in the speed and accuracy of sequencing along with a dramatic reduction in the cost of nucleic acid sequencers. The advances in nucleic acid sequencing have been dramatic, and several disrupting changes in platforms have accelerated the rate at which nucleic acids can be sequenced. Such an event occurred in the early 2000s when next generation sequencing instruments were developed. When compared to the standard measure for the advancement of computational power (CPU, Moore's Law), the capacity to sequence nucleic acids has progressed even faster.

Perhaps the single biggest discovery in the field of nucleic acid sequence technologies was the PCR. It was invented in the mid–1980s and quickly reduced to practice by scientists at Cetus Corporation. Simplistic in principle, PCR can quickly amplify a targeted nucleic acid sequence and create millions of copies of that sequence within a matter of hours (and now minutes). From the initial reagents and instruments for thermocycling, an entire industry arose that quickly developed automated, quantitative PCR-based diagnostic assays. The clinical diagnostics field embraced the technology again, but it was much slower to be adopted for food applications. A recurrent theme of challenges to manual handling, the inhibition of PCR by food components, and the alignment of the method with regulatory standards were encountered and slowly addressed. Today, fully automated instrument is available on the market from Roka Biosciences, the result of a modest transformation of an existing instrument, which was used for clinical samples. This automated instrument still requires preculture enrichment to reach the sensitivity mandated for regulatory purposes.

A number of nonbiochemical, nonnucleic acid techniques are employed. The most common platform is based upon antibody reagents and the presence of unique antigens on the surface of microorganisms. Borrowing from the clinical diagnostics area again, immunoassays are commonly used for the detection and characterization of microorganisms. Formats for immunoassays vary from the iconic microtiter plate enzyme-linked immunosorbent assays to point-of-care dip sticks with a simple visual readout. All of these methods are confirmatory and are used after a presumptive positive is obtained from culture-based methods. In addition, immunoassays can be used for toxin detection as exemplified by the detection of Staphylococcal enterotoxin.

New technological platforms are being explored to improve the sensitivity and impart an increase in specificity. The challenge in most detection technology, including the detection of food-borne pathogens, is something familiar to anyone who has worked with any analytical instrument-signal to noise. There are various signals that might be used to detect a particular organism, traditionally these have been biochemical, then immunological, and more recently genetic. Attempts to harness spectroscopic and other more broad-based physical analytical techniques have been reported (e.g., those based upon mass). The challenge for spectroscopic measurements is that they rely upon an oblique signal that is hard to correlate to a particular pathogen and/or there is a large amount of noise in biological samples. These methods, while they work well when the organism is interrogated in isolation (and better in a vacuum), are challenged when that organism is surrounded by the biological matrix we call "food".

The biggest challenge for the field is the reduction to practical usage of the detection techniques across the entire spectrum of foods. Unlike clinical samples, which are limited in their complexity (stools, blood, tissue biopsy, and cerebral spinal fluid), foods are far more variable with respect to their composition and other microflora. The development of nanoscale methods for detection needs to be interfaced with the macroscopic samples. Finally, these methods need to align with regulatory standards which are slow to evolve to accommodate new definitions of microorganisms and viability. For example, the current US regulations for Listeria monocytogenes in "ready-to-eat" foods suggest potential methodologies for monitoring the organism. These methodologies are all culture dependent. Hence, as technology evolves to eliminate culturing, it will face a challenge in aligning with the regulations unless the regulations also evolve at a similar pace.

Vocabulary

appealing　吸引人
predate　早于
ignorance　无知
fermented　发酵
lenses　镜头
array　一系列
illusive　迷惑人的
taxonomy　分类学
clinical　临床
recombinant　重组
manipulation　操纵
interdisciplinary　跨学科
disrupting　扰乱
platform　平台
thermocycling　热循环
arose　出现

regulatory　监管
modest　适度的
immunoassay　免疫测定
microtiter　微量滴定
immunosorbent　免疫吸附
dip　浸
stick　棒
confirmatory　确认
presumptive　假定
harness　利用
spectroscopic　光谱
oblique　斜
reduction　减少
nanoscale　纳米级
evolve　进化
viability　生存能力

参考译文

食品微生物检测

　　微生物在食品生产中的应用可以追溯到许多世纪,甚至早于微生物世界的发现。尽管对微生物的存在一无所知,但许多发酵食品已经成为不同文化饮食的一部分,它们使用的是配方,但不了解导致这种转化的潜在原因。

　　微生物的检测和鉴定技术大约有100年历史,发生在两个方面。首先是微生物世界的可视化,这在很大程度上归功于安东尼·范·列文虎克的作品,他手工制作的放大镜让他看到了微米级的物体。其次是发现了生物体的培养方法,主要是19世纪80年代罗伯特·科赫的工作。后者的发现使得一些单个微生物在实验室中被分离和生长。从那时起,已经形成了大量不同的生长培养基,并且根据特定的培养条件,现在可以从食物中培养大量的微生物。然而,没有一种方法可以在食物中培养所有微生物,并且有些证明是假的(如弯曲杆菌),从而导致严重低估了被培养的食源性疾病的数量。

在大约75年的时间里,微生物通过基于培养的方法检测,并且它们的鉴定以及整个分类学主要基于显微分析和生化测试。这些技术中的许多技术都是从临床诊断领域衍变而来的,那些领域的样本通常比食物更均匀,更简单。20世纪40年代,Avery,MacLeod和McCarty发现了核酸,但是使用核酸作为常规诊断工具将等到该技术进步到可行的时候。两项技术的发展改变了微生物诊断的格局——核酸测序和聚合酶链式反应(PCR)。它们都是从重组DNA技术领域演化而来,并且建立了体外操作核酸的方法,这些方法又与这些重组核酸分子转化宿主组织的能力相结合。

过去30年来,核酸序列分析已经从一种人工的、费时费力的、需要数天才能完成非常短的序列,演变到以自动化、仪器驱动的过程,它可以在几小时内完成一个完整微生物基因组的测序。从手动过程到自动过程的转变是跨学科工作的结果,这些跨学科工作利用生物学家和工程师将基于酶的核酸序列分析整合到仪器中。第一个自动化的DNA测序仪是在20世纪80年代由Applied Biosystems推出的,随后,一场技术革命迅速导致了测序速度和准确度的指数级增长,同时大幅降低了核酸测序仪成本。核酸测序技术进步十分迅速,平台的几次破坏性变化加速了核酸的测序速度。这个事件发生在21世纪初,当时开发了下一代测序仪器。与提高计算能力的衡量标准(CPU,摩尔定律)相比,核酸序列的进步速度甚至更快。

也许在核酸序列技术领域中最大的一个发现就是PCR。它是在20世纪80年代中期发明的,并迅速被Cetus公司的科学家应用。原理上很简单,PCR可以快速扩增靶向核酸序列,并且在几小时内(现在几分钟)产生数百万个该序列的拷贝。从最初的热循环试剂和仪器开始,整个行业开始迅速发展自动化、基于定量PCR的诊断分析方法。临床诊断领域再一次采用了该技术,但其在食品中应用要慢得多。人工操作经常遇到并在缓慢解决的是,食品成分对PCR的抑制以及方法与国标的一致性问题。现在,Roka生物科技公司在市场上推出了一种全自动化仪器,这是现有仪器适度改造的结果,它被用于临床样品。这种自动化仪器仍需要预培养才能达到监管要求的灵敏度。

许多非生化、非核酸技术被采用。最常见的平台是基于抗体试剂和微生物表面存在的独特抗原。再次从临床诊断领域借鉴,免疫分析通常用于检测和鉴定微生物。免疫分析的方法各不相同,从标志性的微孔板酶联免疫吸附法到使用简单的目测读数的即时检测浸渍棒法。所有这些方法都是验证性的,并且在从基于培养的方法获得假定的阳性后使用。此外,免疫分析还可用于毒素检测,如葡萄球菌肠毒素的检测。

目前正在探索新的技术平台,以提高灵敏度并增加特异性。包括检测食源性病原体在内大多数检测技术面临的挑战,对于使用分析仪器的人来说都很熟悉,信噪比。有许多信号可能用于检测某种特定的生物体,传统上这些信号是生物化学的,然后是免疫学的,最近是遗传学的。已有报道尝试利用光谱学和其他更广泛的物理分析技术(例如基于质量的技术)。光谱测量的挑战在于它们依赖于一个与特定病原体难以关联的倾斜信号和/或生物样品中存在大量噪声。这些方法虽然在机体被隔离(在真空中更好)时运作良好,但当生物体被我们称为"食物"的生物基质包围时,它们就会受到挑战。

该领域面临的最大挑战是如何将检测技术应用到整个食品领域中。与其复杂性有限的临床样品(大便、血液、组织活检和脑脊液)不同,食物在组成和其他微生物菌群方面的变化要大得多。纳米级检测方法的发展需要与宏观样品结合。最后,这些方法需要与缓慢发展的监管标准相一致,以适应新的微生物定义和生存能力。例如,目前美国对"即食食品"中单核细胞增生李斯特氏菌的规定提出了监测生物体的潜在方法。这些方法都是依赖培养基的。因此,随着技术的发展,培养基的使用不断减少,除非法规也以相同的速度发展,否则它将面临与法规不一致的挑战。

Exercise

1. Answer questions

(1) Who invent the magnifying lenses allowed him to see things on the order of a few microns?

(2) When was the nucleic acids discovered?

(3) What maybe the biggest discovery in the field of nucleic acids sequence technologies?

(4) Why is it difficult to practical usage of the detection techniques across the entire spectrum of foods?

2. Translation

(1) The latter development allowed some individual microorganisms to be isolated and grown in the laboratory.

(2) 有许多信号可能用于检测某种特定的生物体,传统上这些信号是生物化学的,然后是免疫学的,最近是遗传学的。

Unit VII
Food Regulation and standard
食品法规与标准

Lesson 31

Reading Material

Food Regulation

Food regulation (or food law) is the complete body of legal texts that establish broad principles for food control in a country, and that governs all aspects of the production, handling, marketing and trade of food as a means to protect consumers against unsafe food and fraudulent practices. In addition, food law should cover the total chain beginning with provisions for animal feed, on-farm controls and early processing through to final distribution and use by the consumer.

Food regulation of China

Generally, Chinese food regulation can be divided into three main levels. The first level constitutes the basic laws: the Food Safety Law, the Agri-food Quality and Safety Law, the Product Quality Law, Agriculture Law, Law of Standardization, Law on the Inspection of Import and Export Commodities. The basic laws are made by the National People's Congress (NPC) which is China's top legislature. The basic laws have the highest legal validity compared with other laws and regulations. The second level is made up by subordinate laws and regulations, such as the Supervision Methods of Quality and Safety in Food Enterprises, Regulation of Food Labels, the Administrative Regulation of Food additives and a great number of national standards establishing food safety requirements. The subordinate laws and regulations are made for guiding and regulating the activity for a specific food, for food production and during trade. The third level consists of various regulations promulgated by provincial governments, for example, the "Administrative measures of local food safety standards of Shanghai". The provincial regulations are made by the provincial governments (or certain municipalities having provincial status) based on the basic and subordinate laws and regulations. The local regulations mainly consist of the measures, detailed rules and regulations for the implementation of controls. The local regulations are the largest in number and the most detailed regulations.

In China, the basic applicable principles of laws and regulations conflicts are: that higher legal ranked laws have priority, new law has priority over an old law, "special" law (i.e. a law applying to a specific food or foods) has priority over a general law. In many of the food safety laws and regulations, the basic laws take priority over the other levels. Among the basic laws, the Food Safety Law is considered to have the highest priority.

Numerous food safety incidents have occurred recently in China, such as the melamine contamination of milk powder, the presence of clenbuterol in pork and plasticizer in beverages, the sale of toxic ginger, and the sale of expired meat produced by Fuxi Company, all of which have cast doubt upon the country's food safety status. As a result, China's food trade and international reputation have been critically affected.

The public concern over food safety has partly prompted the revision of the 2009 Food Safety Law of China (FSL), which was passed by the 14th Session of the 12th Standing Committee of the National People's Congress of China on April 24, 2015. With the process of reform, the 2009 Food Safety Law of China was followed by three rounds of revision and the final amendment is finished in April 2015 with enactment to take place in October 2015. The new FSL contains 10 chapters and 154 articles. The 2015 amendment addresses several issues including online retailing, infant formula, and harsher penalties. This new law that is widely considered one of the most comprehensive and severe pieces of legislation on food safety thus far. The main contents of the 2015 Food Safety Law of China as shown in table 31.1.

Table 31.1　Main contents of the 2015 Food Safety Law of China

Food Safety Law of China	Chapter 1: General Principles
	Chapter 2: Food Safety Risk Surveillance and Assessment
	Chapter 3: Food Safety Standards
	Chapter 4: Food Production and Trading
	Chapter 5: Food Testing
	Chapter 6: Food Import and Export
	Chapter 7: Handling of Food Safety Incidents
	Chapter 8: Supervision and Administration
	Chapter 9: Legal Liabilities
	Chapter 10: Supplementary Provisions

Brief introduction of International Food Regulation

With the increasing of international trade in food, it is essential to have at least a general understanding of international food regulation. International initiatives to coordinate international food regulation and facilitate trade can be divided into three categories: cooperation, mutual recognition, and harmonization. Informal cooperation has existed for many years in a variety of forms. For example, in 1999 the United States and the European Community signed the "Agreement between the United States of America and the European Community on Sanitary Measures to Protect Public and Animal Health in Trade in Live Animals and Animal Products". This agreement covers a wide range of foods (all of animal origin), such as milk and dairy products, seafood, honey, wild game, snails, and frog legs. Harmonization of food regulatory standards has played the most prominent role in recent efforts to facilitate trade.

In 1963, the World Health Organization (WHO) and Food and Agriculture Organization (FAO) create international Codex Alimentarius Commission (CAC) together. The purpose is to protect the health of consumers and maintain the economic interests of the consumers, and guarantee to carry out the food of fair trade and coordinating all the formulation of the standard. CAC now has 173 member states and 1 member organization (EU), the CAC has secretariat (Roma) coordinating committee, executive committee, 6 district coordinating commitlees, 21 professional committee and a government task force.

The World Trade Organization (WTO) clearly stipulates that CAC code standard has a criterion in food trade, and the settlement of international trade disputes is based on the CAC standard or the conclusion of risk analysis. The CAC standard is the basis for promoting international trade and solving international trade disputes, as well as a legitimate weapon to protect the interests of member states.

Vocabulary

amendment　修正案,改善,改正
coordinate　合作
facilitate　促进,使便利
harmonization　和谐,协调
legislation　立法,法律
National People's Congress　全国人民代表大会
nontariff　非关税
standardization　标准化
supervision　监督,管理
subordinate　下级的,附属的

参考译文

食品法规

食品法规(或食品法)是一个国家为了食品控制而建立的具有广泛原则的法律文本,它规定了食品生产、加工、销售和贸易的各个方面,旨在保护消费者免受不安全的食品和欺诈行为的影响。此外,食品法规应当涵盖食品在整个生产过程中的链条,包括动物饲料、农场控制和早期加工一直到消费者最终分配和使用的规定。

中国的食品法规

一般来说,中国的食品法律可以分为三个主要层次。第一个层次由基本法律构成:《中华人民共和国食品安全法》《中华人民共和国农产品质量安全法》《中华人民共和国产品质量法》《中华人民共和国农业法》《中华人民共和国标准化法》《中华人民共和国进出口商品检验法》,这些基本法是由中国最高立法机关全国人大(NPC)制定的,与其他法律法规相比,基本法律具有最高的法律效力。第二层次由附属的法律法规组成,例如《食品生产加工企业质量安全监督管理办法》《食品标识管理规定》《食品添加剂管理条例》以及建立食品安全体系所要求的一系列国家标准,这些法律法规的制定是为了指导和规范特殊食品等的生产和贸易活动。第三级由省级政府颁布的各种地方性法规组成,如"上海市地方食品安全标准的实施办法",省级法规由省级政府(或直辖市)根据基本法律和附属法律法规制定,这些地方法规主要由实施办法、细则和规章组成,地方性法规的数量最多,内容最详细。

在中国,法律法规冲突的适用原则是:上位法优于下位法,新的法律优先于旧法,特别规定的法律(如法律应用到一个特定的食品或一类食品)优先于一般规定,在众多食品安全法律法规中,基本法律优于其他层次法律。而在基本的法律中,《中华人民共和国食品安全法》具有最高优先权。

最近在中国发生了许多食品安全事件:如三聚氰胺污染的乳粉,猪肉中存在瘦肉精,塑化剂饮料,市售有毒生姜和福喜公司生产的过期肉,所有这些事件使人们对国家的食品安全状况感到担忧,中国的食品贸易及国际声誉也受到严重影响。

公众对食品安全的担忧,在一定程度上促使了对2009年颁布的《中华人民共和国食品安全法》的修订,修订版于2015年4月24日经第十二届中华人民共和国全国人大常委会第十四次会议审议通过。修订过程中,2009年的《中华人民共和国食品安全法》历经三轮修订,最终修订案在2015年4月颁布,并于2015年10月生效,新的《中华人民共和国食品安全法》包含10章共154条。2015年的修订版涉及更多问题,包括网上零售、婴儿配方乳粉以及严厉的处罚措施,这部新法律被广泛认为是迄今为止最全面、最严厉的《中华人民共和国食品安全法》。其主要内容如表31.1所示。

表31.1　　2015年颁布的《中华人民共和国食品安全法》主要内容

中华人民共和国食品安全法	第一章:总则
	第二章:食品安全风险监测与评估
	第三章:食品安全标准
	第四章:食品生产经营
	第五章:食品检验
	第六章:食品进出口
	第七章:食品安全事故处置
	第八章:监督管理
	第九章:法律责任
	第十章:附则

国际食品法规简述

随着国际食品贸易量的增加,对国际食品法规需要有一定的了解。当前协调国际食品法规和促进贸易的国际举措可分为三类:合作、相互承认和协调,这些非正式合作已经以多种形式存在多年。如1999年,美国和欧洲共同体签署了《美国和欧洲共同体关于在活体动物和动物产品贸易中保护公众和动物健康的卫生措施的协定》。这项协定所涵盖的食品范围广泛(包括所有动物来源食品),如乳和乳制品、海产品、蜂蜜、野味、蜗牛、青蛙腿。这种食品法规标准的协调在积极促进贸易中发挥了极为突出的作用。

1963年,世界卫生组织(WHO)和联合国粮农组织(FAO)共同成立国际食品法典委员会(CAC),旨在保护消费者健康及维护消费者经济利益,保证开展公正的食品贸易和协调所有食品标准的制定工作。CAC 现拥有173个成员国和1个成员国组织(欧盟),CAC下设秘书处(罗马)、执行委员会、6个地区协调委员会,21个专业委员会和1个政府间特别工作组。

世界贸易组织(WTO)明确规定 CAC 法典标准在食品贸易中具有准绳作用,解决国际贸易争端皆以 CAC 标准或风险分析的结论为基础。CAC 标准成为促进国际贸易和解决国际贸易纷争的依据,也成为保护成员国利益的合法武器。

Exercise

1. Answer questions

(1) What is the definition of food regulation?

(2) What is the main content of China's Food Safety Law?

2. Translation

(1) In China, the basic applicable principles of laws and regulations conflicts are: that higher legal ranked laws have priority, new law has priority over an old law, "special" law (i.e. a law applying to a specific food or foods) has priority over a general law. In many of the food safety laws and regulations, the basic laws take priority over the other levels. Among the basic laws, the Food Safety Law is considered to have the highest priority.

(2) In 1963, the World Health Organization (WHO) and Food and Agriculture Organization (FAO) together to create international Codex Alimentarius Commission (CAC), purpose is to protect the health of consumers and maintain the economic interests of the consumers, guarantee to carry out the food of fair trade and coordinating all the formulation of the standard.

Lesson 32

Reading Material

Food Standard

Standards are used in all realms of human activity in order to specify the characteristics of a product, or its manufacture. In the process, they fulfil a range of functions, such as lowering risks, increasing trust and facilitating predictability in a given market. Standards reduce information costs for market players, which in turn allows for a more efficient functioning of the market.

Food standards are a necessity to both consumers and food industries. They maintain the nutritional values and the general quality of a large part of the national food supply. Without standards, different foods would have the same names, and the same foods different names, both situations confusing and misleading consumers and creating unfair competition.

For international trade in foodstuffs, harmonization of the wide variety of food standards is essential in order to facilitate the global food sourcing trend. As traditional market access barriers are dismantled, nontariff measures offer a tool for the potential protection of domestic products, thus calling for effective forms of food governance.

Food Standard of China

Chinese food standards consist of national standards, professional standards, provincial standards, and company standards. The multi level standards sometimes conflict with each other, and most of those standards are outdated or lower than the international ones. Furthermore, given situations such as the conflicting standards set by different bodies, the regulation based on outdated standards or without established standards, it is necessary to set up the general principles and requirements with regard to the standard setting procedures in order to ensure uniformity and consistence.

An innovation of the new Food Safety Law is the unification of the National Food Safety Standards (the Standards) system. To systematize those food standards, the Food Safety Law entrusts the health administration department of the State Council and the food and drug administration department with the responsibility of consolidating, formulating and publishing the Standards. There are strict procedures for the formulation of national food safety standards. The national food safety standards should be approved by the national committee on standard review which consists of experts with food related background and government officials. In addition to this, before decisions are taken, the committee should consider the comments of different stakeholders and the results of risk assessment as a basis for setting standards. Now, there were more than 1800 national food safety standards and more than 2900 sectoral standards for the food industry in China.

International Food Standards

The trend toward globalization includes growing international trade in food and agricultural products. But, different food laws and standards by various countries impose barriers to trade and raise transaction costs. Increasing trade magnifies the burden of these trade barriers. With growing importance of international trade, it comes a greater need for a uniform reference point in standards.

At the same time, there was a growing recognition of the need to ensure the safety of food. Minimum international standards were needed to protect consumers all over the world. The globalization of ingredient supply chains means that inferior standards in one country may end up adversely impacting the multi-ingredient foods sold in another country.

The FAO and the WHO have the lead in the core elements of standard setting: risk assessment and risk management. Meanwhile, the WTO's laws encourage the design and adoption of international standards by imposing their use as a basis for regulation. The WTO plays a major role in application of food standards in trade and dispute resolution. This circumstance in turn creates a presumption of compliance with WTO rules that ultimately facilitates international trade. International standards thus act as a benchmark for better forms of governance while avoiding the adoption of trade-distorting measures and promoting harmonization.

The CAC has taken a lead role in the international standards for trade in food. At present, Codex comprises more than 200 standards, close to 50 food hygiene and technological codes of practice, approximately 65 guidelines, more than 1000 food additives and contaminants evaluations, and more than 3200 maximum residue limits for pesticides and veterinary drugs.

Vocabulary

standard　标准
foodstuff　食品
innovation　改革
consolidating　统一，巩固
formulating　制订

globalization　全球化
inferior　低等的
publishing　发布
hygiene　卫生

参考译文

食品标准

标准应用于人类活动的所有领域,旨在说明产品或其加工特性。在这一过程中,标准起到了一系列作用,如降低风险、增加信任和促进特定市场的可预测性。标准降低了市场参与者的信息成本,从而使市场能够更有效地运作。

对于消费者和食品工业来说,食品标准是非常必要的,它们确保了国家大部分粮食的营养价值和质量。标准的缺失,可能会导致不同的食品用相同的名称,相同的食品用不同的名称,这两种情况都会给消费者造成困扰和误导,并引起不公平的竞争。

对于国际食品贸易来说,协调各种食品标准是至关重要的,可促进全球食品的采购。随着传统的市场准入壁垒被拆除,非关税措施为国内产品的潜在保护提供了一个工具,因此需要有效的食品管理形式。

中国食品标准

中国的食品标准包括国家标准、行业标准、地方标准和企业标准。这几种多级别的标准有时相互冲突,并且大多数标准已经过时或低于国际标准。此外,鉴于不同机构制定的标准时有冲突、一些标准过时或标准缺失等情况的存在,很有必要制定标准制定程序的一般原则和要求,以确保标准的统一性和一致性。

新《中华人民共和国食品安全法》的一项改革就是对国家食品安全标准(标准)体系的统一。为使这些食品标准系统化,《中华人民共和国食品安全法》委托国务院卫生行政部门同食品药品监管部门来统一、制订和发布这些标准。食品安全国家标准的制订是有严格程序的,国家食品安全标准要由具有食品相关背景的专家和政府官员组成的国家标准审查委员会批准。此外,在做出决定之前,委员会应考虑不同利益攸关方的意见和风险评估结果,以此作为制定标准的基础。目前,我国食品行业已有1800多种食品安全国家标准和2900多种行业标准。

国际食品标准

在全球化趋势下,食品和农产品国际贸易不断增长。但是,各国食品法规和标准的不同对食品贸易造成了障碍,提高了交易成本。贸易的增长又放大了这些贸易壁垒的负担。随着国际贸易的重要性凸显,迫切需要统一标准的参考点。

同时,随着人们对食品安全必要性认识的提高,需要最低的食品国际标准来保障世界各地的消费者。原料供应链的全球化意味着一个国家的劣质标准最终可能会对另一个国家在售的多成分食品产生不利影响。

联合国粮农组织(FAO)和世界卫生组织(WHO)在标准制定的核心要素——风险评估和风险管理中处于领先地位。同时,世界贸易组织(WTO)的法律鼓励制定和采用国际标准,并将其作为监管的基础,WTO利用食品标准在解决贸易争端方面发挥了重要作用,这种情况反过来促进了对WTO规则的遵守,最终促进了国际贸易。国际标准因此成为一种良好监管形式的基准,同时避免采取扭曲市场的贸易措施,促进了协调。

食品法典委员会(CAC)在国际食品贸易标准中发挥了主导作用。目前,该法典包括200多项标准、近50项食品卫生和技术操作规程、约65项准则、1000多项食品添加剂和污染物评价以及3200多项农药和兽药最高残留限量标准。

Exercise

1. Answer questions

(1) What are the categories of Chinese food standards?

(2) Which international organizations have played a important role in the international standards for trade in food?

2. Translation

(1) Standards are used in all realms of human activity in order to specify the characteristics of a product, or its manufacture. In the process, they fulfil a range of functions, such as lowering risks, increasing trust and facilitating predictability in a given market. Standards reduce information costs for market players, which in turn

allows for a more efficient functioning of the market.

(2) The Food and Agriculture Organization (FAO) and the World Health Organization (WHO) have the lead in the core elements of standard setting: risk assessment and risk management. Meanwhile, World Trade Organization (WTO) laws encourage the design and adoption of international standards by imposing their use as a basis for regulation.

Unit VII
Functional Food
功能食品

Lesson 33

Reading Material

Functional Food

There is widespread recognition that diet plays an important role in the incidence of many diseases. Whereas basic nutrients, including vitamins and minerals, are important for growth and development, the focus of functional foods is to provide health benefits beyond those provided by basic nutrients. Although the mechanisms are not completely clear, when eaten on a regular basis as part of a varied diet, functional foods may lower the risk of developing diseases such as cancer or heart disease.

A food can be regarded as "functional" if it is adequately demonstrated to affect beneficially one or more target functions in the body, beyond adequate nutritional effects, in a way that is relevant to an improved state of health. Functional foods should generally be whole natural foods; however, when bioactive substances are extracted from plant or animal tissues, concentrated and added back to food, the resultant functional food may be called a nutraceutical. Functional foods enriched with vitamins, dietary fibres or specific fatty acids or foods that are designed to be low in sodium or saturated fat, can therefore make a valuable contribution to diet.

Several thousands of physiologically active compounds have been identified in functional foods. Each functional food has a different mixture of these active components, which usually are responsible for giving the food its aroma, flavor and color. Several categories of physiologically active components from plant and animal sources were mainly phytochemicals (e. g. terpenes, carotenoids, polyphenols, phytosterols) and zoochemicals (e. g. omega-3 fatty acids-DHA, conjugated linoleic acid). The concentration in food of these components may vary depending on the plant or animal variety, maturity and growth location. Moreover, environmental conditions, such as storage, sunlight, processing, and cooking, may affect the chemical nature, bioactivity and bioavailability of the many compounds in foods.

Some examples of functional components

Carotenoids   Carotenoids are a class of natural fat-soluble pigments found principally in plants, algae and photosynthetic bacteria. They are responsible for many of the red, orange and yellow hues of plant leaves, fruits and flowers, as well as the colors of some birds, insects, fsh and crustaceans. Some familiar examples of carotenoid coloration are the oranges of carrots and citrus fruits, the reds of peppers and tomatoes, and the pinks of flamingoes and salmon. From the plant products commonly consumed by humans, more than 600 different carotenoids (the number includes cis-trans isomeric forms) have been isolated to date.

The most widely studied and well-understood nutritional role for carotenoids is their provitamin A activity. Carotenoids also play an important potential role in human health by acting as biological antioxidants, protecting cells and tissues from the damaging effects of free radicals and singlet oxygen. Lycopene, the hydrocarbon carotenoid that gives tomatoes their red color, is particularly effective at quenching the destructive potential of singlet oxygen. Other health benefits of carotenoids that may be related to their antioxidative potential include enhancement of immune system function, protection from sunburn, and inhibition of the development of certain types of cancers.

Saponins   Saponins are found primarily in legumes, with the greatest concentration occurring in soybeans. Recent experimental investigations suggest that saponins have cholesterol lowering, anticancer and immunostimulatory properties. Anticancer properties of saponins appear to be the result of antioxidant effects, immune modulation and regulation of cell proliferation.

Polyphenols   Thousands of molecules, having a polyphenol structure, (i.e., several hydroxyl groups on aromatic rings) have been identified in higher plants, and several hundred are found in edible plants. In general, phenolic compounds behaving as antioxidants are multifunctional, achieving bioactivity in several ways: fighting free radicals by donating a hydrogen atom from a hydroxyl group (OH) of their aromatic structure; chelating transition metals, such as the Fe^{2+} and Cu^+; interrupting the propagation reaction of free radicals in lipid oxidation; modifying the redox potential of the medium and repairing the damage in molecules attacked by free radicals.

Conjugated Linoleic Acid   Conjugated linoleic acid (CLA) is a mixture of structurally similar forms of linoleic acid and occurs particularly in large quantities in dairy products and foods derived from ruminant animals. The inhibition of mammary carcinogenesis in animals is the most extensively documented physiological effect of CLA, and there is also emerging evidence that CLA may decrease body fat in humans, and increase bone density in animal models.

Brief of functional food products

In the broad history of the East, many plants, such as medicinal herbs, have been used for thousands of years to maintain health and treat diseases. Now, these same ancient remedies are experiencing renewed importance and should be reassessed in our modern age for possible use in the development of high-quality natural health products and dietary supplements for the twenty-first century.

More recently, food companies have taken further steps to develop food products that offer multiple health benefits in a single food. Functional foods have been developed in virtually all food categories. The functional food products can be classified into the following categories. (1) "add good to your life", e.g. improve the regular stomach and colon functions (pre-and probiotics) or "improve children's life" by supporting their learning capability and behavior. However it is difficult to find good biomarkers for cognitive, behavioral and psychological functions. (2) of functional food is designed for reducing an existing health risk problem such as high cholesterol or high blood pressure. (3) consists of those products, which makes your life easier.

The twenty-first century will be an era of new scientific horizons based on the twin forces of globalization and information-intensive industries. We are entering a period of unparalleled opportunities and intense competition.

Vocabulary

bioactivity   生物活性
carotenoid   类胡萝卜素
chelating   螯合
functional   功能性的
linoleic acid   亚油酸

lycopene   番茄红素
phytosterol   植物甾醇
polyphenol   多酚
saponin   皂苷
terpenes   萜烯

参考译文

功能性食品

人们普遍认为饮食在很多疾病中能发挥重要作用，其中的基本营养物质如维生素和矿物质对机体的生长发育非常重要，而功能性食品的重点则是提供基本营养物质外的对身体有益的功能。虽然这些机制尚不完全清楚，但作为日常饮食的一部分，功能性食品可以降低患癌症或心脏病等疾病的风险。

如果一种食品除了适当的营养作用外，还对身体有一种或多种有益功能，在一定程度上能改进健康状态，那这个食品可被视为具有"功能性"。功能性食品一般为全天然食品；然而，从动植物组织中提取、浓缩和重新添加到食品中而具备功能性的食品称为保健食品。富含维生素、膳食纤维、特定脂肪酸、低钠或低饱和脂肪的功能食品在饮食中起着举足轻重的作用。

食品中有生理活性的功能性成分已鉴定出几千种。每种功能食品由几种不同活性成分混合而成，它们通常对食品的香气、风味、色泽起作用。从动植物中提取的这些有生理活性的功能成分可分为植物化学物（如萜烯类、类胡萝卜素、多酚类、植物甾醇）和动物化学物（如 ω-3 脂肪酸——DHA、共轭亚油酸）。食品中这些活性成分的浓度可能取决于动植物种类、成熟度和生长位置。此外，环境条件如储存、光照、加工和烹调也可能影响食品中许多化合物的化学性质、生物活性和生物利用度。

功能性成分举例

类胡萝卜素　类胡萝卜素是一类天然的脂溶性色素，主要存在于植物、藻类和光合细菌中。许多植物的叶、果实和花中的红色、橙色、黄色与类胡萝卜素有关，还有些鸟类、昆虫、鱼和甲壳动物的色泽也与它有关。常见的与类胡萝卜素颜色有关的例子有：胡萝卜和柠檬的橙色、辣椒和番茄的红色、红鹳和三文鱼的粉红色。目前从人们普遍食用的植物产品中，已分离出超过 600 种不同的类胡萝卜素（包括顺反式异构体）。

对类胡萝卜素研究最广泛和对其营养作用有深入了解的是其作为维生素 A 的前体物质的活性。类胡萝卜素在人体健康中起着潜在的重要作用，包括生物抗氧化剂的作用、保护细胞和组织免受自由基和单线态氧的破坏。番茄红素是一种能使番茄呈红色的类胡萝卜素，能有效地淬灭单线态氧的潜在破坏力。其他类胡萝卜素对身体的有益功能可能与它们潜在的抗氧化能力有关，包括增强免疫系统功能、防止皮肤晒伤和抑制某些癌症的扩散。

皂苷类 皂苷主要存在于豆类中，其中大豆里面含量最高。最新实验研究表明皂苷能降低胆固醇、抗癌和增强免疫特性。皂苷的抗癌能力似乎与其抗氧化作用、免疫调节作用及细胞增殖调控作用有关。

多酚 高等植物中已发现上千种具有多酚结构（芳香环上含多羟基）的化合物，其中从可食用的植物中发现有几百种。一般而言，酚类化合物通过抗氧化特性来实现多种生物活性：通过酚羟基上的一个氢原子与自由基反应；与过渡金属离子螯合，如 Fe^{2+} and Cu^+；中断脂质氧化过程中自由基的链式反应；改变介质的氧化还原电位并修复受自由基攻击的分子。

共轭亚油酸 共轭亚油酸（CLA）是一种结构相似的亚油酸的混合物，尤其在乳制品和反刍动物源食品中大量存在。有趣的是，烹饪或加工后 CLA 反倒增加。在动物中 CLA 能有效抑制乳腺癌的发生，也有新的证据表明 CLA 可以降低人体脂肪，动物模型中它可增加骨密度。

功能性产品简介

在东方悠久的历史长河中，利用许多植物如草药来维持健康和治疗疾病已有数千年。现在，这些古老的疗法正重新获得重视，并重新评估其开发为现代天然优质健康产品和膳食补充剂的作用。

当前，食品公司已大力开发在单一食品中提供多种健康益处的产品，几乎所有类型的食品都开发有功能食品。功能食品可以分为以下几种：①"让你的生活更好"，如改善肠胃功能（益生菌和益生素）或通过支持儿童的学习能力和行为来"提高儿童能力"，当然找到一种对认知、行为和心理功能进行标记的物质是困难的；②开发专为减少现有健康风险如高胆固醇和高血压的功能食品，如高胆固醇或血压高；③让你的生活变得更简单的产品。

二十一世纪将是一个基于全球化和信息密集化的新科学时代。我们正在进入一个前所未有的机遇和挑战并存的时期。

Exercise

1. Answer questions
(1) What are the functional foods?
(2) List some common functional components and their role.

2. Translation
(1) A food can be regarded as "functional" if it is adequately demonstrated to affect beneficially one or more target functions in the body, beyond adequate nutritional effects, in a way that is relevant to an improved state of health.

(2) Several thousands of physiologically active compounds have been identified in functional foods. Each functional food has a different mixture of these active components, which usually are responsible for giving the food its aroma, flavor and color.

Related glossary terms

abrasive 研磨剂,研磨的
accelerate 加速
acid/base 酸/碱
actin 肌动蛋白
actomyosin 肌动球蛋白
adulterants 杂物,混合物
agent 药剂
aleurone 糊粉
algal 藻
allergens 过敏原
ambivalent 矛盾的
amendment 修正案,改善,改正
amino acid 氨基酸
amorphous 无定形的
amylopectin 支链淀粉
amylose 直链淀粉
anthocyanin 花青素
antioxidant 抗氧化剂
appealing 吸引人
arable 耕地
aroma 芳香,香味
arose 出现
array 一系列
artificial 人工
ascorbate 抗坏血酸盐
ascorbic acid 抗坏血酸(维生素C)
aspartame 天冬氨酰苯丙氨酸甲酯
atomic absorption 原子吸收
atrazine 阿特拉津(一种除草剂名)
attenuated 衰减的过去式
automatable 自动化的
bacteria 细菌
baked 烘焙,烤
batch 一批

bentonite 膨润土;皂土
biosynthesis 生物合成
bitter 苦的,苦味
bonding 成键
bran 糠,麸
breakage 破坏,破损,破损量
brinjal leaf 茄子叶
broken 碎米(粒)
brown rice 糙米
brush machine 刷米机
bulgur 碾碎的干小麦
burgeoning 迅速增长的
butter 黄油
Butylated Hydroxy Anisole (BHA) 丁基羟基茴香醚
Butylated Hydroxy Toluene (BHT) 丁基羟基甲苯
calcium 钙
canning 罐头制作、灌装,罐头加工的
caramel coloring 焦糖色素
carbamate 氨基甲酸酯
carbaryl 胺甲萘
carbohydrate 碳水化合物
carmine 胭脂红
carotenoid 类胡萝卜素
cartridge 管筒
casein 酪蛋白
cellular 细胞的
celluloses 纤维素
centrifugation 离心作用
cereal 谷类植物
chamber 室,膛
characteristic 特征、特性、特点
cheese 奶酪

chelate　螯合
chlorinate　使氯发生作用,用氯消毒
chromatography　色谱分析
chromium　铬
clinical　临床
coagulate　凝结
coagulum　凝结物,凝固物
coarse　粗糙的,粗鄙的
coconut　椰子
cofactor　辅因子
collagen　胶原质,胶原蛋白
colloidal　胶体
colorant　着色剂
column　柱
comminute　弄碎,把……弄成粉末
communication　交流,沟通
compatible　兼容
complementary　互补
components　成分,组分
components　组件
comprehensive　综合
comprise　包含
conditioners　调节剂
confectionaries　糖果,蜜饯
confirmatory　确认
consistent　一致性
consolidating　统一,巩固
contaminant　污染物
contaminate　污染,弄脏
conventional　传统的
coordinate　合作
copper　铜
cordials　浓缩果汁,饮品
cripple　削弱
crumb　面包心;面包屑
crystal-clear　透明似水晶的,易懂的
crystalline　水晶般的
curcumin　姜黄素
curd　凝乳

customer　顾客
customers　顾客(复数)
dairy　乳制品
decay　腐烂
decontamination　消除污染,净化
defect　缺点,缺陷
de-hulling　脱壳
dehydrate　脱水
dehydrated　脱水的,干燥的
dehydrogenase　脱氢酶
denaturation　变性
denature　使变性;使变质
deprivation　损失,丧失
derivatization　衍生化
detached　分开的,分离的
deviation　偏差
differential　差别的,特定的,微分的
dilute　稀释,冲淡
dip　浸
diphenylamine　二苯胺
disaccharides　双糖,二糖
discharge　释放;排出,释放
discoloration　变色
disinfestation　灭虫
disintegration　瓦解;蜕变;崩溃;〈物〉裂变
disrupting　扰乱
document　文件,记录
dose　剂量
dough　(用于制面包和糕点的)生面团
dunst　粗粉
dye　染料,染色
electrostatic　静电
emission　发射
emulsifier　乳化剂
emulsion　乳液,乳剂
enamel　搪瓷
erythorbate　异抗坏血酸盐(用于食物中作为抗氧化剂)
establish　建立

eventually 终于；最后
evolve 进化
excitation 激发
extend 延长,延伸
facilitate 促进,使便利
fat 脂肪
fatsoluble 脂溶性
favor 风味
fermented 发酵
fertilizer 化肥
filtration 过滤
flavonoid 类黄酮
flavoring agent 风味物质
flesh 肉；肉体；果肉
fluorescence 荧光
fluorimetry 荧光测定法
food additive 食品添加剂
foodstuff 食品
formulating 制订
fourier 傅里叶
fragments 片段（复数）
free radical 自由基
fructose 果糖
gastrointestinal 胃与肠的
gel 凝胶；胶化
gelatin 凝胶,白明胶
germ 幼芽,胚芽,胚原基
gliadin 麦胶蛋白
globalization 全球化
glucose 葡萄糖
gluten 面筋
glutenin 麦谷蛋白
grind 磨（碎）；碾（碎）
guar gum 瓜尔豆胶
guava 番石榴
guideline 指导方针,指南
harmonization 和谐,协调
harness 利用
hazard 危害

headspace 顶部空间
high-speed horizontal mixer 高速卧式搅拌机
homeostasis 动态平衡
homogenization 均质
homogenizer 均质器,高速搅拌器
hood 风帽,头巾；机罩
hopper （磨粉机等的）漏斗,送料斗,加料斗
hopper 加料斗；送料斗；漏斗
horizontal 地平的；水平的
hormone 激素
hybridization 杂交
hydration 水合,水合作用
hydrolysis 水解
hygiene 卫生
hyphenated 连接
ignorance 无知
illusive 迷惑人的
immunoaffinity column 免疫亲和柱
immunoassay 免疫测定
immunosorbent 免疫吸附
impart 传递、传授
implementation 实施,履行
implicated 有牵连的
impurity 杂质,混杂物
incubation 孵化；孵育
inferior 低等的
infrared spectroscopy 红外光谱法
innovation 改革
inorganic 无机的
instruction 指令,使用说明
integrity 完整
interdisciplinary 跨学科
interfere 干扰,干涉,妨碍
intestinal 肠道的
iodine 碘
iron 铁
irradiation 照射,辐射

issue 问题
jam 果酱
jellies 果胶,果冻(jelly 的复数)
kidney 肾脏
kieselgur 硅藻土
knead 揉;按摩;捏(面团、湿黏土等);揉捏(肌肉等)
lactalbumin 乳清蛋白
lactobacillus bulgaricus 保加利亚乳杆菌
lactoglobulin 乳球蛋白
lamb 羔羊肉;羔羊,小羊
lard 猪油
lead 铅
legislation 立法,法律
lenses 镜头
lipid 脂质,脂类
lipoprotein 脂蛋白
liquidiser （通常指电动的）果汁机
loaf 一条(块)面包
localised 局限性的
locust-bean gum 刺槐豆胶
lubricate 涂油
lukewarm （液体）微温的
macrophage 巨噬细胞
magnesium 镁
magnetic nanoparticles 磁性纳米颗粒
magnetic 有磁性的
management 管理,管理人员,管理部门
mandatory 强制性的
manganese 锰
manipulation 操纵
manufacture 制造,生产
matrix 基质
mechanisms 机制,机理
melamine 三聚氰胺
mesh 网眼,筛孔,网络
methodology 方法
mettwurst 生熏软质香肠
micelle 胶束,胶囊
microfiltration 微量过滤
microflora 微生物群落
microorganism 微生物
microtiter 微量滴定
middlings 皮磨细粒,细麸
mill 磨坊;制造厂
mineral 矿物质
mint 薄荷
modest 适度的
molybdenum 钼
monascus red 红曲红
monosaccharides 单糖
mould 霉菌
multiply 乘、繁殖;多样的
mycotoxins 真菌毒素
myosin 肌球蛋白
naked 裸露的
nanoscale 纳米级
National People's Congress 全国人民代表大会
natural antioxidant 天然抗氧化剂
natural colorant 天然着色剂
natural flavoring substance 天然风味物质
nectar 果肉饮料;花蜜
neurological 神经上的
neurotoxic 神经毒性
niacin 烟酸
noble 贵族(的)
nontariff 非关税
nuclear 原子核
nucleoprotein 核蛋白
objectionable 令人不快的,令人反感的,讨厌的
oblique 斜
oil 油
olive 橄榄
optimum 最适宜(的)
organophosphate 有机磷
overtail 筛上物

palatability 适口性
passion fruit 西番莲果,百香果
pasteurization 巴氏杀菌法
pasteurize 用巴氏灭菌法对(牛乳等)消毒
（灭菌）
pathogen 病原体
pathogenic bacteria 病菌,病原菌
pathogenic 致病性
peanut 花生
peeling 碾,削
pectic 果胶的,黏胶质的
pectin 胶质
pectolytic enzyme 果胶分解酶
pepper 辣椒;胡椒粉
perforate 打孔
perishable 易腐败的
pesticide 农药,杀虫剂
pharmaceutical 药物的
phenolic 酚类
phosphorus 磷
phytochemicals 植物化学物质
phytopathogen 植物病原体
plant 工厂
platform 平台
pneumatic conveyor 气动输送机
poisoning 中毒
polished rice 精米
polished 擦亮的,磨光,精练的
polyethylene 聚乙烯
polypeptide 多肽
polyphenol 多酚
pork 猪肉
portable 便携式
potassium 钾
poultry 〈集合词〉家禽
preceding 在前的;在先的
precursor 前体
predates 早于
prerequisite 先决条件,首要必备的

preservation 保藏,保存
preservative 防腐剂
presumptive 假定
primers 引物
principle 原则
probable 可能的
projecting 突出的,伸出的
promulgation 颁布
Propyl Gallate (PG) 没食子酸丙酯
protein 蛋白质
protocol 协议
proving period 醒发阶段
publishing 发布
pulper 碎浆机;搅碎机;(咖啡豆的)果肉采集器
qualitative 定性
quality 质量
quantification 定量
rack 支架;架子
raman 拉曼
recombinant 重组
reconstitution 重新组建,重新构建
record 记录
rectangular 长方形
reduction 减少
refrigeration 制冷、冷藏
regeneration 再生
regulate 监管,控制;调整
regulatory 调节的
rehydrate 再水化,水合
render 致使
reoccurrence 复发,重现
residue 残渣
residues 残留,剩余
resonance 共振
retailer 零售商
retained 保留的
review 检查,复审
rice milling 碾米

ridge　背脊,山脊
ripening　成熟
rough rice　毛谷、毛稻
repair　修复
rugged　坚固的
saccharin　糖精
saucepan　长柄而有盖子的深平底锅,炖锅
scattering　散射
scrape　刮,擦
screening　筛选
seasoning　调味品,佐料
seed　种子
selenium　硒
semolina　砂子粉,皮磨粗粒
senescence　衰老
sers　表面增强拉曼光谱法
settle　沉淀,澄清
shaft　轴,转轴
shearing force　剪力
shelf life　保存期,货架期
shelling　去壳,去皮,砻谷
shoe　喂料器
shortening　起酥油
shorts　细麸
sieve　筛,滤网
silicon　硅
smell　嗅觉
smelting　冶炼
sodium benzoate　苯甲酸钠
sodium copper chlorophyllin　叶绿素铜钠盐
sodium　钠
solubility　溶解度
solution　溶液,溶解
sophisticated　复杂的
sorbent　吸附剂
sorbic acid　山梨酸
sour　酸味
soxhlet　索氏提取
spectroscopic　光谱

spectroscopy　光谱学
spice　调味料,香料
spoilage　腐败,变坏
squash　果汁汽水;南瓜小果
stabilizer　稳定剂
standard　标准
standardization　标准化
starch　淀粉
stationary　静止的,不动的
sterilization　消毒,杀菌
sticks　棒(stick复数形式)
straightforward　简单的
straw　稻草,麦秆,茎秆
streptococcus thermophilus　嗜热链球菌
strip-based immunoassay　试纸免疫测定
subcritical　亚界
subordinate　下级的,附属的
substitutes　替代品(substitute复数形式)
sucrose　蔗糖
suction　吸入,吸力,抽气,抽气机,抽水泵,吸引
supercritical　超临界
supervision　监督,管理
supplement　补充剂
suspension　悬挂,悬浮液
sweetener　甜味剂
symptoms　症状(symptom复数形式)
synergistic interactions　协同增效作用
synthesis　合成
synthetic antioxidant　合成抗氧化剂
synthetic colorant　合成着色剂
tandem　串联
tart　尖刻的;酸的
taste　味觉
taxonomy　分类学
tedious　单调乏味的
temper　调和,调节
temperamental　气质的;性格的;不可靠的
tenderize　使嫩化、使软化

Tertiary Butyl Hydroquinone（TBHQ） 叔丁基对苯二酚
textural 质构化,纹理性
thermocycling 热循环
thiabendazole 噻苯咪唑
thiamin 硫胺素(维生素B_1)
thresh 打谷,打,颠簸 vt. 脱粒,翻滚
tissue 组织
titration 滴定法
TLC 薄层色谱
tocopherol 生育酚
trace 痕迹,踪迹,微量
transgenic 转基因
transition 过渡
umami 鲜味
unanimously 全体一致地
utensil 器具、用具、器皿
utilized 利用的过去式
valve 阀,阀门;电子管
vanilla 香草;香草精

vapor 蒸气压
veal 牛肉;小牛
vegetative insecticidal protein（Vip） 营养期杀虫蛋白
viability 生存能力
vibrate 颤动
vibrational 振动
virulence 毒性
viscoelastic 黏弹性
viscolizer 均质机,匀化器
viscosity 黏稠,黏性
vitamin 维生素
volatile 挥发性
voluntary 自愿的
water activity 水分活度
water 水,水分
whey 乳清
whipping 搅打……变稠
yeast 酵母
zinc 锌

References

[1] 许学勤. 食品专业英语文选(第二版)[M]. 北京：中国轻工业出版社, 2016.

[2] 汪洪涛, 陈宝宏, 陈成. 食品专业英语(第二版)[M]. 北京：中国轻工业出版社, 2016.

[3] 许学书, 谢静莉. 食品专业英语[M]. 北京：化学工业出版社, 2008.

[4] 庞杰, 刘先义. 食品质量管理学[M]. 北京：中国轻工业出版社, 2017.

[5] 邹良明. 食品仪器分析[M]. 北京：科学出版社, 2012.

[6] 张建新, 陈宗道. 食品标准与法规[M]. 北京：中国轻工业出版社, 2017.

[7] 张晓燕. 食品安全与质量管理(第二版)[M]. 北京：化学工业出版社, 2010.

[8] Gutiérrez-López G F, Barbosa-Cánovas G V. Food Science and Food Biotechnology[M]. Boca Raton FL: CRC Press, 2003.

[9] Mudambi S R, Rao S M, Rajagopal M V. Food Science[M]. New Delhi: New Age International (P) Limited, Publishers, 2006.

[10] Brennan J G. Food Processing Handbook[M]. Weinheim: Wiley-VCH, 2006.

[11] Bartosz G. Food Oxidants and Antioxidants Chemical, Biological, and Functional Properties[M]. Boca Raton FL: CRC Press Taylor & Francis group, 2014.

[12] Belitz H D, Grosch W, Schieberle P. Food Chemistry[M]. Berlin: Springer, 2009.

[13] Newton D E. Food Chemistry[M]. New York: Facts On File, 2007.

[14] Nielsen S S. Food Analysis[M]. New York: Springer, 2010.

[15] Chen X D, Mujumdar A S. Drying Technologies in Food Processing[M]. West Sussex UK: Blackwell Publishing Ltd, 2008.

[16] Eskin N A M, Shahidi F. Biochemistry of Foods[M]. London: Elsevier

Inc, 2013.

[17] Yildiz F. Advances in Food Biochemistry[M]. Boca Raton FL: CRC Press Taylor & Francis group, 2009.

[18] Henry J. Advances in Food and Nutrition Research[M]. London: Elsevier Inc, 2013.